Rethink, Retool, Reboot

Praise for this book

'Technology is always about politics and social justice: who wins, who loses, and what directions are chosen. This book puts these themes centre-stage in a clear and accessible style. Anyone interested in how technology can work for people, justice, sustainability and development should read it.'

Ian Scoones, Director ESRC STEPS Centre,
Institute of Development Studies, University of Sussex

'This is an important book with an important message at a perfect time as the international development community struggles to get to grips with technology and innovation. Many of the answers to many of the development challenges are already out there, yet in our drive to showcase our innovative thinking we often turn our backs on them in favour of more complex, "smarter", "innovative" solutions. Yet it is often the simpler ones that work best. "It is time", as Simon reminds us, "to reboot our relationship with technology". We need a people-first and problem-first approach at a time when many are technology-first. This book makes a hugely valuable contribution to the debate, and should be essential reading for technologists and policy makers alike.'

Ken Banks, Founder of kiwanja .net,
and National Geographic Emerging Explorer

'*Rethink, Retool, Reboot* is a valuable accompaniment to the 2015 Sustainable Development Goals which combine a commitment to eradicate poverty and to manage our planet sustainably. Trace explains and reinforces the massive role that technology plays in this existential mission. But he highlights in a clear and accessible way how we need to revisit who controls it, who benefits and who misses out if we are not to make our world even more unequal and do more damage to our planet. He brings alive the relationship between technology, the environment, poverty, and power in a way that anyone interested in people and the planet will be stimulated by and learn from. '

Mark Goldring, Chief Executive, Oxfam Great Britain

'Technology is the great enabler, but uncontrolled use of technology can also pose a threat to humanity, as our experience of fossil fuels and climate change demonstrates. This book explores two crises – why so many people in the world are still without access to technologies necessary for even the most basic standard of living, and why the bulk of technology innovation fails not only to address their needs, but also brings the threat of environmental change that hangs over us all. A compelling argument to change the way we govern the development and use of technology, and a recommended read for anyone interested in technology and justice.'

Jeremy Leggett, Founder of Solarcentury and SolarAid,
Chairman of Carbon Tracker

'Amidst the frenzy of global interest in digitization and robotization, this book brings the focus of technology back to where it is most needed: empowering the world's poorest communities. Simon Trace's concept of Technology Justice is fresh, powerful and much needed.'

Kate Raworth, Environmental Change Institute, Oxford University

'If ever there was a time to re-think the role of technology, this has to be it. With a renewed emphasis on the Sustainable Development Goals and on climate change, technology is at the heart of all our hopes and many of our fears for the future. Covering all today's "big issues", Simon Trace provides a fresh and eloquent approach to innovation, governance issues and access to technology, compellingly brought together through this rallying call for Technology Justice.'

Jonathon Porritt, Founder Director, Forum for the Future

'Simon Trace takes his inspiration from Schumacher's *Small is Beautiful* but extends the metaphor of magnitude. Through careful use of data, case study and analysis he argues that scale is not in itself beautiful, but is the key to sustainability and impact. In reflecting on scale, on how small an intervention needs to be or how large it can conceivably be, Trace gets to the heart of technological innovation for development: it's all about perspective. Trace shows us that if technology justice is about one thing it's about bringing together perspectives that matter – historical, contextual, technological, public, private and communal – to shape scalable, sustainable innovation. Trace's book is in some respects a 21st century reboot of Schumacher's work and a hugely important contribution to our thinking around technology for development.'

James Smith, Professor of African & Development Studies,
University of Edinburgh

'Technology should never be considered as an ultimatum, that just because we can do something, we should. Even Winston Churchill who was fascinated by and obsessed with science and technology once famously said that it should be "on tap, but not on top". Simon Trace issues a comprehensive invitation to rethink what we ask of technology. Public debate is full of reports hypnotized by technological novelty and innovation for its own sake. But Trace reveals how low-tech solutions can often outperform high-tech ones – delivering multiple benefits to the people who need them most. *Rethink, Retool, Reboot* says it is time to move on, and critically assess each technology – whether it is the product of a small farmers workshop or a giant corporation's laboratory – to find what will really help us all thrive within planetary boundaries.'

Andrew Simms, co-director New Weather Institute,
author Cancel the Apocalypse

About the author

Simon Trace is a consultant and writer on technology and international development. A chartered engineer with an MA in the anthropology of development, Simon has over 30 years of experience working on technology in relation to energy, water, food and natural resource management across South Asia, Sub-Saharan Africa and Latin America. Simon's career has been spent principally in international development NGOs, notably 20 years with WaterAid, including five years as its International Director and, most recently, 10 years as the CEO of Practical Action. Simon has also spent time under secondment to Unicef, the Agricultural Development Bank of Nepal and CARE, and has held international advisory roles including membership of Steering Group of the United Nations Sustainable Energy for All initiative's Global Tracking Framework and membership of the external advisory group for the World Bank's Readiness of Investment in Renewable Energy index (RISE). He is a trustee of the European Environment Foundation and was awarded a CBE for services to international development in 2016.

Rethink, Retool, Reboot

Technology as if people and planet mattered

Simon Trace

Practical Action Publishing Ltd
The Schumacher Centre,
Bourton on Dunsmore, Rugby,
Warwickshire, CV23 9QZ, UK
www.practicalactionpublishing.org

A catalogue record for this book is available from the British Library.
A catalogue record for this book has been requested from the Library of Congress.

ISBN 978-1-85339-904-6 Hardback
ISBN 978-1-85339-905-3 Paperback
ISBN 978-1-78044-904-3 Library Ebook
ISBN 978-1-78044-905-0 Ebook

Citation: Trace, S., (2016) *Rethink, Retool, Reboot: Technology as if people and planet mattered* Rugby, UK: Practical Action Publishing, <http://dx.doi.org/10.3362/ 9781780449043>

Since 1974, Practical Action Publishing has published and disseminated books and information in support of international development work throughout the world. Practical Action Publishing is a trading name of Practical Action Publishing Ltd (Company Reg. No. 1159018), the wholly owned publishing company of Practical Action. Practical Action Publishing trades only in support of its parent charity objectives and any profits are covenanted back to Practical Action (Charity Reg. No. 247257, Group VAT Registration No. 880 9924 76).

Cover design by Mercer Design
Indexed by Elizabeth Ball
Typeset by Allzone Digital Services Limited
Printed by Short Run Press, UK

Contents

Acknowledgements xii

Acronyms xiii

Prologue 1

Introduction 3

Part I: Rethink 11

1. **Defining technology and technological progress 13**
   Defining technology 13
   The idea of technological progress 14
   The social dimensions of technology 21
2. **Technology Justice: establishing the principle 25**
   Technology and environment 25
   Technology and human wellbeing 28
   A safe operating space for humanity 36
   Defining a principle of Technology Justice 38
3. **Technology Justice and access to basic services 43**
   The light bulb – will it ever catch on? Access to energy services 43
   The tap – a technology whose time has finally come? 53
   Critical yet unavailable: access to essential medicines 60
   Technology Justice and basic services 66
4. **Technology Justice and access to knowledge 71**
   Feeding the world: smallholder farmers' need for better access to technical knowledge 71
   The digital divide 77
   Technology Justice and the social foundation 85
5. **Technology Justice and use 91**
   Justice as compromise 91
   Industrialized agriculture and biodiversity loss 92
   Energy security and climate change 98
   The misuse of antibiotics: turning the clock back for medicine 105
   Technology Justice and use 111
6. **The governance of technology access and technology use: time for a rethink 117**
   Injustices in access to and use of technology 117
   A single unified problem 118

Part II: Retool: Driving innovation to develop the right technologies 121

7. The link between technological innovation and economic development 123
Justice as a fair space for innovation 123
Understanding how innovation happens 124
Innovation systems and developing economies 130

8. Technology Justice and innovation systems in practice 137
Justice and the management of risk in innovation 137
Justice and the shaping of the purpose of innovation 142
Driving sustainable pro-poor innovation: the need for change 154

9. Intellectual property rights: part of the solution or part of the problem? 161
Why patent? 161
Do patents encourage innovation? 162
TRIPs, patents, and the negative impacts on developing countries 163
Asymmetries of power 165
Alternatives to the existing patent system 166
Where next? 171

10. Recognizing the role of the state in effective innovation systems 175
Venture capital and the valley of death in the energy sector 176
Recognizing reality: governments engage in entrepreneurial activity 182
Changing the narrative: rebalancing expectations of public and private-sector roles 185

11. Beyond market forces: other drivers for innovation 189
Innovation for what? 189
Innovation for whom? 193
Innovation by whom? 196

12. Making technology innovation work for people and planet: the need to retool 209

Part III: Reboot: Building a different approach to the governance of technology 213

13. Reimagining technology as if people and planet mattered 215
The need to reboot our relationship with technology 215
Rebooting access 218
Rebooting use 223
Rebooting innovation 229
Changing the way technology is governed 231
Technology as if people and planet mattered 234

Epilogue: Is small beautiful? 239

Appendices 243

Figures, tables, and boxes

Figures

1.1	Headlines for news feeds on 'Technology'	13
2.1	Rockström's safe planetary boundaries	27
2.2	Proportion of global population meeting minimum social foundation	34
2.3	The role of technology in establishing a safe, inclusive, and sustainable space for development	37
3.1	Top 20 electricity access deficit countries in proportion and absolute population terms	45
3.2	Indicative levelized costs of electricity for on-grid and off-grid technologies in sub-Saharan Africa, 2013	47
3.3	Average annual investment in access to electricity, by type, needed to meet universal access by 2030	48
3.4	Lending for off-grid energy access by the MDBs compared to the IEA requirement to ensure universal access by 2030	48
3.5	Kinshasa electricity access measured by connections and by energy services	51
3.6	Proportion of population by region with access to improved drinking-water supplies or sanitation, 2012	55
3.7	Inequalities between populations that practise open defecation in Mozambique	55
3.8	Disproportionate share of DALY burden from tuberculosis, HIV/AIDS, and malaria in Africa compared to other regions	61
3.9	Availability of selected medicines in public and private health facilities, 2001–07	63
4.1	Relative global size and number of different types of agricultural extension programme	74
4.2	Percentage increase in economic growth per 10 per cent increase in penetration of telecommunication technologies	82
4.3	The energy–food–water nexus	88
5.1	Proportion of global anthropogenic greenhouse gas emissions by source	99
5.2	High and low-emission scenarios and impact on global temperature rise	100
5.3	Predicted global annual death rates resulting from antimicrobial resistance compared to other diseases in 2015	106
5.4	The geographical spread of morbidity in 2050	107
7.1	The production curve	126

7.2 The impact of technology innovation on the production curve 126
7.3 Connecticut's innovation system 129
7.4 Stages in the development of an innovation system 133
8.1 Sources of funding for R&D on neglected diseases, 2009 144
8.2 Attainable agricultural production per unit of land or person as a function of resource investment (capital and labour) 151
8.3 Global expenditure on agricultural R&D, 2000 153
10.1 The clean-energy innovation cycle and valleys of death 177
10.2 Energy-sector technology innovation projects and types of finance available 178
10.3 The origins of technology used in Apple's iPod, iPhone, and iPad 184
13.1 Technology Justice – a safe, inclusive, and sustainable space for humankind 216
13.2 The Sustainable Development Goals 219
13.3 Outline of a circular economy 228

Tables

2.1 Expected consequences of crossing planetary boundaries and the technological challenges of staying within them 29
2.2 Elements of a social foundation and proportion of global population below this floor 33
2.3 Technologies required to support Raworth's social foundation 35
3.1 Energy services required for a small health post 46
3.2 Multitier matrix for access to household electricity supply 50
3.3 Water and sanitation policy implementation of pro-poor governance, monitoring, and finance targets 57
3.4 Availability of health services around the world 62
5.1 Severity of potential impacts of global temperature rise 100
8.1 Technologies introduced in the last century and their unexpected negative impacts 138
8.2 Common concepts used in debates on the precautionary principle and appropriate actions 141
8.3 Distribution of global health R&D funding across neglected diseases, 2011 146
8.4 Examples of yield gaps for key crops in developing-world regions 150
8.5 Agricultural R&D investment options outlined in the IAASTD Global Report 152
9.1 Comparing alternative innovation incentive systems 168
10.1 Investment and targeted returns for a typical US venture capital fund portfolio 177
11.1 Potential lines of questioning for responsible innovation governance approaches 190

Boxes

5.1 Agroecological approaches that build resilience 96
8.1 International treaties that contain references to the application of the precautionary principle 140
11.1 Different levels of 'inclusivity' of innovation 194
11.2 Open-source definition: distribution terms open-source software must abide by 198

Acknowledgements

I would like to thank Practical Action for providing me with the inspiration, time, and space to write this book.

Particular thanks are also due to Toby Milner for the general support and advice he offered to me throughout this project as well as his hugely helpful editorial work. I remain very grateful for his patience and calm in the face of a project that took somewhat longer than originally anticipated!

I'd also like to extend sincere thanks to those who took the time to review and comment on parts or all of the draft text, notably Astrid Walker Bourne and Jonathan Casey from Practical Action; Dr Jamie Cross, Edinburgh University; Dr David Grimshaw, Royal Holloway, University of London; Kate Raworth, Oxford University; Professor Ian Scoones, Institute of Development Studies; Andrew Scott, Overseas Development Institute; and Dr Peter Cruddas, Cranfield University. Their feedback was immensely helpful.

Finally, I would like to recognize Dr Vishaka Hidellage and the then staff of Practical Action's programme in Sri Lanka who, back in 2011, first coined the phrase Technology Justice and started the process that eventually led to this book.

Permissions

Figures 2.2, 2.3, 13.1 (pages 34, 37, 216 respectively) have been adapted by the publisher from *A Safe and Just Space for Humanity: can we live within the doughtnut?* (2012) with the permission of Oxfam GB www.oxfam.org.uk. Oxfam does not necessarily endorse the text that accompanies these figures, nor has it approved the adaptation.

Acronyms

AEIC	American Energy Innovation Council
AGRA	Alliance for a Green Revolution in Africa
AIDS	see HIV/AIDS
ARPA	Advanced Research Projects Agency of the US Department of Defence
AMC	advanced market commitment
BSE	bovine spongiform encephalopathy
CGIAR	Consortium of Agricultural Research Centres
COP	Conference of Parties of the United Nations Framework Convention on Climate Change
DALY	disability-adjusted life years
EC	European Commission
EU	European Union
FAO	Food and Agriculture Organization of the United Nations
FDA	United States Food and Drug Administration
FDI	foreign direct investments
FIT	feed-in tariff
GATT	General Agreement on Tariffs and Trade
GAVI	Global Alliance for Vaccines and Immunisation
GDP	gross domestic product
GEF	Global Environment Facility
GFRAS	Global Forum for Rural Advisory Services
GHG	greenhouse gas
GLAAS	Global Analysis and Assessment of Sanitation and Drinking-Water
GPTs	general-purpose technologies
GSMA	association that represents the interests of mobile operators worldwide
HBN	Honey Bee Network
HIV/AIDS	human immunodeficiency virus/ acquired immune deficiency syndrome
IAASTD	International Assessment of Agricultural knowledge, Science and Technology for Development
ICTs	information and communication technologies
IDRC	International Development Research Centre, Canada
IEA	International Energy Agency
IFAD	International Fund for Agricultural Development
IFC	International Finance Corporation
IMF	International Monetary Fund

INDC	intended nationally determined contribution
IPCC	Intergovernmental Panel on Climate Change
IPR	intellectual property rights
IRR	internal rate of return
ITU	International Telecommunications Union
LDC	least developed countries
LED	light-emitting diode
MDB	multilateral development bank
MDGs	Millennium Development Goals
MOOC	Massive Open Online Course
NFCTSAR	National Federation for the Conservation of Traditional Seeds and Agricultural Resources
NGO	non-government organization
NIS	national innovation systems
NME	new molecular entity
NWSC	Nairobi Water and Sewerage Company
OECD	Organisation for Economic Cooperation and Development
OSDD	Open Source Drug Discovery initiative
OSM	Open Source Malaria
OSSI	Open Source Seed Initiative
PSM	People's Science Movement
PTD	participatory technology development
REACH	Registration, Evaluation, Authorisation and Restriction of Chemicals
R&D	research and development
RML	Reuters Market Light
RRDI	Sri Lankan Rice Research and Development Institute
RRI	responsible research and innovation
SDGs	Sustainable Development Goals
SDI	Slum Dwellers International
SGC	Structural Genomics Consortium
SPRU	Science Policy Research Unit
STN	Social Technologies Network
TAC	Treatment Action Campaign
TFM	Technology Facilitation Mechanism
TRIPS	Trade-Related Aspects of Intellectual Property Rights
UN	United Nations
UNDP	United Nations Development Programme
UNEP	United Nations Environment Programme
UNFCCC	United Nations Framework Convention on Climate Change
SE4All	United Nations Sustainable Energy for All
USDA	United States Department of Agriculture
WHO	World Health Organization of the United Nations
WIPO	World Intellectual Property Organization
WTO	World Trade Organization

Prologue

The origins of this book are personal. My background is in engineering and I have spent over 30 years in international development, working mainly on water and sanitation, soil and water conservation, and energy. As a result, I have built up a very practical interest in the factors that affect technology innovation and technology dissemination. But the primary inspiration for this book comes from my 10-year association with the international development organization Practical Action and the work and thinking of its founder, the economist E.F. Schumacher.

Schumacher was the author of *Small is Beautiful* (1973), a book that, alongside Rachel Carson's *The Silent Spring* (1962), helped to inspire the environmental movement in the 1960s and 1970s. *Small is Beautiful* starts with the argument that the traditional discourse on economics is fundamentally flawed, based as it is on an idea that development relies on perpetual economic growth which, in turn, depends on ever-increasing consumption of material resources. Schumacher introduced the concept of 'natural capital', talked about the finiteness of natural resources, and used the field of energy to demonstrate how the consumption patterns of Europe and North America could never be replicated on a global scale. His conclusion was that humanity was on a collision course with nature and needed to take action quickly.

Surprisingly quickly, just over 40 years since the publication of *Small is Beautiful*, we now seem to be on the cusp of that collision. What has got us here is not quite the exhaustion of resources that Schumacher envisioned. Instead, as climate change indicates, we have managed to choke ourselves on the pollution from the burning of fossil fuels before the fuels themselves have actually run out. Nonetheless, we are faced with incontrovertible evidence that Schumacher's warning was right and that humankind cannot continue to exist on a model of living that prioritizes ever-increasing consumption over everything else.

This is a major problem when 40 per cent of the world's population have to exist on less than $2 a day (UN, 2015a) and require significant increases in their levels of consumption to reach a reasonable minimum standard of living. It is an even bigger problem when one considers the predicted growth of the world's population from 7.3 billion in 2015 to 9.7 billion by 2050 (UN, 2015b) and the additional consumption of resources that will accompany this growth.

Schumacher's approach to economics in *Small is Beautiful* covered a lot of ground, including environmental sustainability, food production, and the purpose of work and education. But at its heart was recognition of the critical role technology plays in delivering material wellbeing and facilitating our interaction with the environment we inhabit. Schumacher argued that our trajectory of technological development was taking us in the wrong direction.

Automation and the technology of the mass production line were stripping meaning from work and breaking important ties that had traditionally bound communities together. The financial cost of creating workplaces with these new technologies was too high to make them applicable to solving issues of poverty in the developing world, and perhaps even to creating full employment in the developed world. Even if this were not the case, the environmental cost of trying to do so would be too great.

Essentially, Schumacher issued a call to rethink technology – what it is for, how it is developed and used, and, most importantly, how universal access to technologies that are critical to a basic standard of living can be ensured without breaking the ecological carrying capacity of our planet. Some 40 years later, parts of *Small is Beautiful* seem dated but the core of Schumacher's thinking is as relevant today as it was when the book was published. That we are in a fight with nature we cannot win, that access to technology is essential to achieving a reasonable minimum standard of living, and that technology choice is critical for humanity to live sustainably on Earth, are all ideas that have stood the test of time.

Inspired by Schumacher, this book takes a fresh look, through the lens of technology, at the twin problems of ending poverty and ensuring an environmentally sustainable future for everyone on this planet. It does not argue for a 'technical fix' for poverty or environmental degradation. Social, political, cultural, and economic factors clearly shape these problems and must be addressed as integral to any solution. This book does, however, argue that ending poverty and achieving environmental sustainability cannot be realized without radical changes to the way technology is developed, accessed, and used. Technology is at once a crucial part of the problem and of the solution. A new form of governance of technology development and use must be found: one that is fairer and more equitable than the present; one that takes the interests of both current and future generations into account. The purpose of this book is to explore what that new governing principle – Technology Justice – could look like.

References

Carson, R. (1962) *The Silent Spring*, Boston, MA: Houghton Mifflin.

Schumacher, E.F. (1973) *Small is Beautiful: A Study of Economics as if People Mattered*, Oxford: Blond and Briggs.

UN (2015a) *Resources for speakers on global issues – vital statistics*. New York: United Nations Resources. <http://www.un.org/en/globalissues/briefingpapers/food/vitalstats.shtml> [accessed 4 December 2015].

UN (2015b) *World Population Prospects – Key findings and advanced tables*. New York: Population Division, Department of Economic and Social Affairs, United Nations.

Introduction

Most aspects of human development have gone hand in hand with technological change: the first stone tools and their implications for hunting and food processing, or the invention of the plough and its impact on agriculture, for instance. We have even named historical epochs after the technologies that are considered to be their key defining feature: the Stone, Bronze and Iron Ages, the Industrial Revolution, the Age of Steam, and now the Information Age. Technological change has continued to expand human potential and, through the invention of such instruments as the telescope and the microscope, to extend what it is possible for humans to know. Technology development and adaptation have enabled people to achieve wellbeing with less effort and drudgery, or at lower cost and with fewer resources. Technical innovation – the use of new knowledge, tools, or systems – continues to be key to humankind's ability to make more effective use of the resources available to us, to respond to social, economic, and environmental changes, and thus to improve our wellbeing. As the sociologist David Nye says: 'it is easy to imagine human beings as preliterate, but it is difficult to imagine them as pre-technological' (Nye, 2006: 5).

More than ever, human wellbeing and the possibility of a sustainable life on this planet are both tightly bound to the technological choices we make. In a world on the cusp of environmental disaster and with nearly 3 billion people living in absolute poverty, the way humanity chooses to govern the development, dissemination, and use of technology is crucial to finding a just, equitable, and sustainable solution for present and future generations. But today that technology governance system, or at least such of it that exists, is broken and unfit for purpose with its failure resulting in great injustices.

It is possible that historians of the future will look back on 2015 as a watershed moment. Hopefully, with hindsight, it will be seen as the year when the international community finally grasped the urgency of the two great challenges of global poverty and climate change and started to make concerted effort on both fronts. Certainly, 2015 provided many opportunities for political leaders to move from empty rhetoric to real action. The year started off with a global gathering in Japan in March to confirm the Sendai Framework – an international agreement to guide progress on disaster risk reduction to 2030 and a successor to the Hyogo Framework for Action. That was followed in July by the third International Conference on Financing Development in Addis Ababa, a meeting that aimed to create a global framework for financing sustainable development for the next 15 years.

However, it was the final two events of the year that stole most of the thunder. In October an ambitious new set of Sustainable Development Goals (SDGs) was agreed at the United Nations in New York. Unlike their

http://dx.doi.org/10.3362/9781780449043.001

predecessors, the Millennium Development Goals (MDGs), which focused largely on improving conditions in the developing world, the SDGs attempt to address environmentally sustainable development for the world as a whole and set out the changes required of developed as well as developing nations. The resulting 17 goals and 169 targets, unsurprisingly, are far more complex than the MDGs, making their agreement all the more remarkable.

Then the Paris Climate Conference (COP21) in December was hailed as a major breakthrough, with national pledges to cut greenhouse gas emissions adding up to something that might just keep global warming within the 'safe' limit of 2°C, an achievement that finally removed the stench of failure that had hung over the climate change talks since their nadir at COP15 in Denmark in 2009.

These latter two events attracted huge and, on balance, positive international media coverage, allowing the year to close with a sense that 'something was being done'. The question is whether that 'something' is enough to ensure historians will indeed look back at 2015 as a pivotal year. Unfortunately, the answer to that question, for now at least, has to be a firm 'no', on two grounds.

Firstly, both the SDGs and the results of COP21 are more aspirational pledges than mutually agreed actionable plans and, as such, have no real consequences for any party that fails to deliver on them. There is no mandatory requirement for national governments to adjust their plans to take account of the SDGs. Indeed, the SDGs, while representing an eminently sensible and necessary set of proposed actions, have been criticized precisely for lacking a theory of change as to how the setting of global goals is supposed to influence national government behaviour and, what is more, for failing to take the opportunity to learn from the experience of the MDGs about what works and what doesn't in this respect (see, for example, Green, 2015). Moreover, neither the national emissions targets (the so-called Intended Nationally Determined Contributions) nor the financial commitments made during the Paris talks are legally binding, and history shows that a lack of binding agreement generally leads to large gaps between expressed intentions and actual outcomes in the United Nations climate talks.[1]

The second reason why the outcome of COP21 and the agreement of the SDGs will not, alone, be enough to place 2015 as the year the world got its act together, is the principal topic of this book – technology.

Access to modern or improved technology is clearly essential to provide adequate food, water, energy, shelter, and livelihoods for everyone on the planet. In the developed world, modern, science-based technology underpins everyday life: food production, access to basic services such as water or electricity, the building materials and energy efficiency of homes, transport infrastructure, the delivery of children's education, basic health services, the ability to communicate, recreation, and personal security and safety. Indeed, its ubiquity renders it close to invisible. In the developing world, by contrast, the absence, for many, of access to improved technologies is stark and almost always a marker of extreme poverty. That absence means not just a life made hugely arduous by the lack of modern mechanical and labour-saving devices, but also a life exposed to unnecessary risks to health. The smoke from cooking over open fires, for example, leads to an annual 4 million premature deaths from respiratory and cardiovascular diseases, and cancer among children and adults (WHO, 2014), while 800,000 children under the age of five die every year from diarrhoeal disease arising from a lack of safe water supplies and safe sanitation (Liu, 2012).

Lack of access to improved technology also means a life made less productive. Time spent collecting water and fuel wood, for example, keeps children, especially girls, out of school and adults, mostly women, away from other productive economic activity. In sub-Saharan Africa the UN estimates women collectively spend some 40 billion hours a year collecting water – the equivalent of a year's worth of labour time by the entire workforce of France (UNIFEM, 2008). At the heart of this contrast is an injustice of monstrous proportions. Nearly 40 per cent of the world's population are denied access to technologies that already exist and which the remaining 60 per cent would consider critical to attaining a minimum standard of living. This book will argue that it is an injustice because it is both unfair and unnecessary, the result of choices that could be made differently and priorities that could be changed.

Yet, unfettered access to technology can create huge problems. Since the Industrial Revolution we have used energy technologies based on fossil fuels (coal, oil, and gas) to promote rapid technological change that has delivered massive positive improvements in the quality of life for billions of people. But that progress has come at a cost. The resulting carbon emitted into the atmosphere is causing global mean temperature to rise, leading to a cascade of negative impacts on food production and the availability of fresh water, increases in the intensity and frequency of natural disasters, and a rise in sea level that threatens low-lying coastal zones in countries such as Bangladesh and the very existence of small island states. Again, this is resulting in great injustice. The use of technology – in this case a suite of technologies based on fossil fuels – is leading to vast and unintended negative consequences for the environment, raising questions about the sustainability of human and other forms of life on this planet. Despite being equipped with this knowledge, we have so far been unable to summon the political will to invest in the vital shift to clean technologies at anywhere near the rate necessary to avoid significant temperature rise and the consequent adverse impacts on current and future generations around the globe to be able to live the lives they value.

One of the achievements of 2015 was that the SDG process firmly recognized that the twin challenges of global poverty eradication and the achievement of environmental sustainability can no longer be dealt with in isolation. Climate change does not respect national boundaries and can only be tackled by a global effort. But developing countries cannot be expected to remain underdeveloped in order to allow richer countries to continue emitting carbon to maintain their higher standard of living. Global collaboration on climate change is only possible if adequate support and attention is paid to reducing carbon footprints in the developed world while, at the same time, assisting the developing world to make rapid progress on improving living standards via a green development path. This understanding is reflected in the final choice of SDG goals and targets and is to be commended.

The SDG and the COP processes also clearly consider the issue of technology as paramount. The use of technology, technology transfer, or technical capacity building is specifically mentioned in the targets for 11 of the 17 SDGs.[2] Moreover, a number of entities have been put in place to deal with issues of technology. The SDG process includes a Technology Facilitation Mechanism and a Technology Bank, while the climate change talks have a Subsidiary Body for Scientific and Technological Advice that oversees a 'Technology Mechanism' made up of a Technology Executive Committee and a Climate Technology Centre and Network.

The reason 2015 cannot yet be seen as a watershed moment is not therefore because of a failure to focus simultaneously on the twin challenges (the SDGs do this). Neither is it because of a failure to realize the importance of technology as part of the solution to these two issues (this is clearly accepted in the SDG and COP21 outcomes). Rather, it is because both the SDG process and the climate talks have ignored a very important fact: the way in which the development and use of technology is governed (or rather the failure to govern it) is leading the world in completely the wrong direction, away from, instead of towards, the achievement of a minimum acceptable and environmentally sustainable standard of living for everyone on the planet now and in the future.

Examples of this failure abound. This book will show that the drivers of technical innovation remain largely market-oriented, meaning technological innovation to alleviate poverty is a low priority because of the low 'purchasing power' at the bottom of the pyramid, despite some loud rhetoric about the commercial opportunity offered by the scale of that segment of the market. In a market system where environmental factors are still considered externalities, there is a similar lack of motivation to accelerate innovation to meet the urgent need for new green technologies. Across most key sectors, few global forums exist to help identify and prioritize areas of technology research or to coordinate global research agendas. Even fewer global mechanisms exist that can then provide finance to fund priority areas. As a result, research efforts are duplicated and resources wasted at a point in history when time is of the essence. Likewise, coordinated global approaches to identify and manage the potential risks associated with the development of new technologies remain largely absent, throwing up barriers to a coherent approach to risk management.

Similar issues restrict access by the poor in the developing world to many existing technologies critical to establishing a basic standard of living, such as water and energy supplies or health services. A lack of voice in national and local planning processes and a lack of purchasing power in commercial markets are perhaps two of the biggest barriers. This, combined with a lack of reliable data on existing levels of access to critical services (particularly in rural areas where people may rely more on 'off-grid' provision of services for water, sanitation, and energy), makes planning harder and holding government to account more difficult. In some sectors a focus on technology as a commodity that can be traded misses the point that more effective solutions might simply involve better access to improved technical knowledge. In the agricultural sector, for example, the Green Revolution technologies of the 1960s (technical packages of fertilizer, hybrid seeds, herbicides, and pesticides, together with irrigation) have failed to benefit large swathes of smallholder farmers across Africa because of a combination of high cost and a general unsuitability of the technologies to the marginal nature of the lands being farmed and the lack of irrigation. The new technological revolution being proposed is more of the same – proprietary packages of bioengineering and genetically modified crops – despite the fact that, health and environmental concerns aside, these technologies are likely to face the same obstacles to effective implementation (affordability and relevance) that were encountered in the Green Revolution solutions for African smallholders. Meanwhile, building knowledge systems and technical skills around low-input agroecological approaches to food production, which might be better able to affordably raise productivity on marginal lands, remains difficult to finance.

Governance issues are not restricted solely to technology innovation or to lack of access in the developing world. Mechanisms to ensure the use of technology by some does not stop others from being able to live the lives they value are also weak or non-existent. The COP process has demonstrated just how difficult it is to wean the world off fossil fuel technology, despite the overwhelming scientific evidence that failure to do so will leave a potentially disastrous legacy for future generations. But at least that is the subject of global political talks through the COP process. By contrast, there is very little public debate and no COP equivalent to mediate the impact of agricultural industrialization on the narrowing of the genetic base of our food systems and the consequent increase in vulnerability to future risks from new pests and diseases or environmental change. Likewise, there is no global process to address the fact that the huge advances in human health afforded by antibiotics are at risk as a result of our inability to control their use – an inability that could bring the age of microbial control to an end less than a century after it commenced.

For 2015 to be seen as a truly watershed moment, the SDG and COP processes have to be capitalized on quickly by following through on these immensely important issues of technology governance, or rather the lack of it. There is a very limited window of opportunity to stabilize the climate before irreversible change results from greenhouse gas emissions. Unprecedented progress has to be made in addressing poverty and inequality in the developing world and re-engineering the technological foundations of societies. That, and the intentions of the SDG and COP processes, cannot be achieved without a radical change in the way technology innovation, dissemination, and use is governed.

This book explores why the governance of technology is so important and what changes need to take place to resolve these issues. It is divided into three sections. Part I starts by looking at notions of technological progress and the relationship between technology and human development before introducing the concept of 'Technology Justice' and demonstrating the need to 'rethink' how we use and provide access to technology. Part II goes on to explore the idea that we need to 'retool' – to re-examine our innovation processes – in order to focus on driving technology development towards, rather than away from, the twin problems of poverty and environmental sustainability. The book closes with a third section that uses the principles of Technology Justice to set out a series of radical changes required to 'reboot' our relationship with technology, a necessary step-change in the journey to finding a sustainable and equitable future for everyone on this planet.

Political, social, cultural, and economic factors determine the way new technologies are developed and used, and shape opportunities for access to technologies essential to achieving a minimum standard of living. It would be pointless to argue that by focusing on technology alone, problems of poverty and environmental degradation could be solved. Nevertheless, the way we choose to develop, use, and provide access to technology will play a critical role in their solution. The core purpose of this book is therefore not to argue that there can be a technical fix for ending poverty and addressing environmental degradation. It is instead to call for the issue of how we govern technology to be treated with the urgency it deserves, as a crucial part of the political, social, and economic change process that will determine whether humanity has a future on Earth.

Notes

1. The Green Climate Fund is a good example of this. Set up under the 2010 climate talks (COP16) in Cancun, Mexico, and intended to be central to obtaining $100 bn a year of climate finance by 2020, the fund had, as of mid-2015, secured just $10 bn of pledges, of which only $5.5 bn had been provided as actual cash (Green Climate Fund, 2015).
2. The use of technology, technology transfer, or technical capacity building is specifically referenced in the targets for SDGs 2, 3, 5, 6, 7, 8, 9, 11, 12, 14 and 17 as per the outcome document of the United Nations Summit for the adoption of the post-2015 development agenda (UN, 2015).

References

Green, D. (2015) 'From Poverty to Power: Hello SDGs, what's your theory of change?', 29 September, *Oxfam blogs*, <http://oxfamblogs.org/fp2p/hello-sdgs-whats-your-theory-of-change/> [accessed 2 December 2015].

Liu, L. et al. (2012) 'Global, regional, and national causes of child mortality: an updated systematic analysis for 2010 with time trends since 2000', *Lancet,* 379: 2151–61 <http://dx.doi.org/10.1016/S0140-6736(12)60560-1>.

Nye, D.E. (2006) *Technology Matters: Questions to Live With*. Cambridge, MA: The MIT Press.

UN (2015) 'Final draft of the outcome document for the UN Summit to adopt the Post-2015 Development Agenda', *Sustainable Development Knowledge Platform*, <https://sustainabledevelopment.un.org/content/documents/7603Final%20draft%20outcome%20document%20UN%20Sept%20Summit%20w%20letter_08072015.pdf> [accessed 16 July 2015].

UNIFEM (2008) *Progress of World's Women: Who Answers to Women? Gender and Accountability*. New York: UNIFEM (now UN Women).

World Health Organization (WHO) (2014) *Household Fuel Combustion: WHO Guidelines for Indoor Air Quality*. Geneva: WHO.

PART I: RETHINK

Why technology is not working for human development or environmental sustainability and why things need to change

CHAPTER 1
Defining technology and technological progress

> *It is primarily through the growth of science and technology that man has acquired those attributes which distinguish him from the animals, which have indeed made it possible for him to become human.*
>
> Arthur Holly Compton

This book is about technology injustice. It shows that the way we govern access to, development, and use of technology is unfair and, ultimately, unsustainable. It is also about how a principle of Technology Justice offers a different way of looking at technology and insight into how technology could be used to create a sustainable and equitable future for everyone. Before those ideas can be explored in depth, it is necessary to clarify what is meant by the term 'technology' in this book. It is also important to describe briefly some of the debates around the notion of what constitutes technological progress, how it occurs, and the social dimensions of technology, in order to understand how conventional views of technological progress as both inevitable and progressive may not be helpful.

Defining technology

The word 'technology' is open to wide interpretation. Today it seems to be most often used to refer to electronic gadgets, mobile phones, and the internet. A quick analysis of the topics of articles on the online technology pages of four major news agencies on the day of writing this chapter (Figure 1.1) supports this assertion.

Figure 1.1 Headlines for news feeds on 'technology'

Source: Word cloud produced from analysis of the topics of all articles on the first pages of the online news pages on 'Technology' from the BBC, CNN, Sky, and the *Telegraph* newspaper on 14 May 2015.

http://dx.doi.org/10.3362/9781780449043.002

This book takes the view that technology extends far beyond this limited field to include traditional indigenous technologies and knowledge as well as the vast array of technology and technical knowledge that underpins the high standard of living achieved in the developed world today.

The *Encyclopaedia Britannica* has the following to say about the origins of the word:

> The term 'technology', a combination of the Greek *technē* ('art' or 'craft') with *logos* ('word' or 'speech'), meant in ancient Greece a discourse on the arts, both fine and applied. When it first appeared in English in the 17th century, it was used to mean a discussion of the applied arts only, and gradually these 'arts' themselves came to be the object of the designation. By the early 20th century, the term embraced a growing range of means, processes, and ideas in addition to tools and machines. By mid-century, technology was defined by such phrases as 'the means or activity by which man seeks to change or manipulate his environment'. (Buchanan, 2014)

As the encyclopaedia notes, such a broad definition fails to distinguish between technological activity and scientific inquiry. It also has the potential to incorporate forms of organization such as political systems and markets. This may be why there has been a narrowing in how the term is generally used, at least in common speech, in recent decades. Use of the term 'high technology' was first noted in English in the early 1960s to refer to the practical applications of modern science and, by the early 1970s, this had been shortened to 'high-tech' (Harper, 2015). It is probably as a result of the usage and connotations of 'high-tech' that the word 'technology' is today more likely to be associated with information and communication technologies – computers, telephones, applications of the internet, and so on – as Figure 1.1 confirms.

In this book, by contrast, the word 'technology' is taken to refer to *the tools, machinery, artefacts, and systems of technical knowledge that humans use to interact with the natural environment and each other*. This encompasses technology based on recent science and, equally, tools, practices, or techniques based on traditional knowledge, for example: a horse-drawn plough, an Archimedes screw,[1] and traditional techniques for the selection and breeding of seeds or the control of soil erosion. To provide some practical limits to the subject, though, the definition used here does not extend to what could be described as non-technical systems of knowledge, such as political or managerial systems and practices.

The idea of technological progress

Technology is often presented in the media and everyday discussion in the abstract – the rational outcome of the application of the latest science to a real-world problem. A common view of the relationship between humans and technology is of a historical and linear progression with humanity constantly inventing and innovating to achieve ever higher levels of wellbeing. This idea that modernization comes about or is evolved through access to ever more sophisticated levels of technology has, together with the concept of economic growth, underpinned ideas of development for the last century.

Reality is a bit messier than this. Technology is a product of human interactions and the use and innovation of technology inevitably reflects the political, social, and cultural nature of the societies from which it emerges. Moreover, human beings shape and, in turn, are themselves shaped by technology. That 'messiness' means that, in reality, technological progress is not as linear or inevitable as we might like to believe and its social impact not as easy to predict as we would wish.

Questioning technological determinism and the inevitability of technological progress

In his book *Science and Technology for Development* the Edinburgh-based academic Professor James Smith traces how views of development through technological progress have changed over the years. In the 1960s, one school of thought saw development in terms of a linear process of modernization, whereby countries pass through a five-stage model from 'traditional society' via industrialization to an 'age of mass consumption' with 'widespread affluence, urbanisation and the consumption of consumer durables'. More recently, the alternative idea of 'technological catch-up', whereby countries can develop their skills base and use new technologies to leapfrog stages of the linear model to catch up or even overtake richer, 'leader' countries, has been something 'that many countries aspire to' (Smith, 2009: 14–17). The mobile phone is often cited as an example of technological leapfrog, with many developing countries virtually abandoning the costly extension of landline services into rural areas in favour of the more flexible and less capital-intensive mobile phone, but the idea could equally apply to leapfrogging whole stages of industrialization.

This vision of constant technological progression can lead to a sense of technological determinism – a belief that certain inventions have within them the seed of an inexorable chain of events. For example, that the efficiency of the wheel must inevitably lead to its universal adoption, that the development of the gun must lead to the abandonment of spears and swords, or that the introduction of the combustion engine must spell the end of horse-drawn transport and ploughs. Many social scientists, however, do not see technological 'progress' as such a linear or inevitable affair. David Nye, for instance, cites three counterfactuals to the above examples:

- The Japanese abandoning guns for cultural reasons after adopting them from Portuguese traders in the middle of the 16th century, not picking them up again until the mid-19th century (the Samurai preferred swords and arrows, which had more symbolic meaning for them).
- The present-day Amish community's rejection of modern transport and agricultural technology in favour of the horse-drawn cart and plough in the USA.

- Evidence that the Mayans and the Aztecs knew of the wheel (they put them on toys and ceremonial objects) but that they did not use wheels in construction or transportation.

Nye goes on to suggest that it is cultural choices rather than any inherent logic or usefulness of particular technologies that determine whether or not they are adopted: 'in short, awareness of particular tools or machines does not automatically force a society to adopt them or to keep them' (Nye, 2006: 18–20).

This view of culture shaping technological choices is echoed more recently in rejections of programmes for childhood vaccination against polio in parts of Nigeria and Pakistan, culminating in the murder of 26 polio workers in Pakistan and 10 in Nigeria in 2013. The reasons behind this rejection appear complex and while, in some cases, doubt over the efficacy of the vaccine itself seems to be part of the rationale for parents refusing to have their children vaccinated, other factors also play a part. These include suspicion of a programme addressing a disease not viewed locally as a health priority and, in areas of conflict and active insurgency, a boycott of the programme being a means to assert power and challenge government authority (Baron and Magone, 2014).

The impact of culture can also be seen in the regular use of multiple fuels and stoves for cooking, known as 'fuel stacking' in the developing world. Many households routinely use two or more fuels. Studies in Latin America show that even households that have switched to liquefied petroleum gas for most of their cooking still rely on less efficient and more polluting stoves or even open fires to cook certain foods, for example the daily staple tortilla. Similar patterns of use have been documented in Asia and Africa. Although some of this behaviour can be linked to household income and the cost of fuel for the improved cook-stoves, cultural issues associated with preferred cooking practices or taste are also cited as reasons for not adopting the cleaner, healthier, and more efficient cook-stoves for all cooking tasks (SE4All, 2013).

Culture can be seen as an influence not only on the behaviour of consumers, determining what technologies are adopted and used, but also on technology producers and investors, determining what technologies are offered for adoption in the first place. Social relationships and ties between producers and financiers in a market, for example, can play a more important role than the efficacy of a technology itself in determining what technologies are brought to market. Jamie Cross's exploration of the history of the development and large-scale uptake of a simple solar lantern manufactured under the brand of 'd.light' is a good example of this (Cross, 2013). The d.light is aimed at poor consumers in the developing world who have no access to mains electricity. It is a low-cost and robust electric light with an LED bulb and a built-in rechargeable battery and solar panel. Founded in 2006, the d.light company had sold over 6 million units by 2014, outstripping most other solar lantern manufacturers in the process (d.light, 2014). According to Cross, the success of the d.light lamp was not due to technical superiority over other similar lamps already on the market in countries such as India. It owed more to the company's ability to raise investments from venture capitalists to scale up its operations. This ability stemmed mainly from two facts:

- The founder members of the company were all alumnae of an 'Entrepreneurial Design for Extreme Affordability' course at Stanford University and were able to trade on their university's name and utilize the extensive social network associated with the institution to gain access to potential investors.
- The company was able to create a compelling narrative around the product for potential investors, largely based on the chief executive's personal experience of living off-grid in Benin for four years as a Peace Corps volunteer and his stories of the hardship and danger that the d.light could help alleviate.

Views of how technological progression happens have changed over time. The idea of societies needing to go through set stages of technological development has given way to the possibility of 'leapfrogging' over stages in certain circumstances to achieve technological catch-up. Notable exceptions have been found to the deterministic view of new technologies inevitably muscling out older and less effective ones, as the examples given here show. The shortcomings of technological determinism in explaining the adoption of new technology are important to note, however, given the influence of that line of thought in early economic theories on the impact of technological progress on growth (which is looked at in more depth in Chapter 6). They are also important to remember when critically reflecting on the rhetoric around emerging technologies today. Research in genomics and nanotechnology, for example, 'has been shown to carry highly optimistic promises of major social and industrial transformation, suggesting a need ... to instil some form of responsibility in disentangling present hype from future reality' (Stilgoe et al., 2013: 1571).

The unpredictability of the social impact of technology

Technological determinism can also take the form of assuming that adoption of a technology will necessarily lead to a certain social outcome or impact. Take as an example the idea that access to the internet will inevitably lead to a more open and democratic society as people are exposed to global news sources and despots can no longer hide the truth from their people. Or that the adoption of latrines as a safe form of sanitation will automatically lead to a reduction in diarrhoeal disease.

Again these ideas often do not stand up to scrutiny. Although it is argued that social media was used to support the bid for greater self-expression and democracy during the Arab Spring in 2011 (Howard et al., 2011), the internet is also being used today to garner support for the formation of an Islamic fundamentalist proto-Caliphate in Iraq and Syria under ISIS (Channel 4, 2014). And while a resident of a slum in Dhaka might build and use a latrine, the prime motivation for doing so may not be health but a desire for increased privacy instead of having to defecate in the open. Such a motivation may not also lead to the adoption of the necessary additional hygiene behaviours (hand-washing after defecation and before eating, hygienic storage of water and food in the home, etc.) that would, alongside safe disposal of faeces in a latrine, result in the prevention of diarrhoeal disease.

Given that technology and culture are inextricably entwined, gendered roles bestowed on men and women by society also play a part in shaping the development and use of technology. This can be seen in the way technology itself has become gendered. It is argued, for example, that during the latter half of the 19th century technology increasingly became associated with the rise of mechanical and civil engineering disciplines – professions dominated by men – thereby diminishing the link with women's knowledge and expertise in the process. Technology became 'male machines rather than female fabrics' (Wajcman, 2009). This inherent masculinity of technology can be viewed as a barrier to both women's access to technology and their power to shape its design and evolution.

The link between gender-ascribed roles in work and the introduction of technology is an important one. The application of new technology can positively affect women's lives by reducing the physical effort or the adverse health impacts of gender-ascribed roles: piped water supplies reducing the burden of water collection or more efficient cook-stoves reducing exposure to harmful indoor air pollution, for example. But not all impacts, even from well-intentioned efforts, are positive for women. The introduction of new technologies can also result in the transfer of tasks or activities from the domestic realm to the commercial realm and lead to a shift of employment opportunities from women to men. In Bangladesh, the husking and polishing of rice by hand was a traditional source of income for women from the poorest rural households who were employed by richer neighbours. With the introduction of mechanical rice mills, not only did better-off households prefer to take their rice to the mill, but also the (fewer) new employment opportunities generated by the mechanization process were filled predominantly by men (Begum, 1983). This is not a new phenomenon. English legal records show that, in the 13th and 14th centuries, women were the chief brewers of ale, the common drink of the rural population. When brewing later became commercialized, control of the process shifted to men (Nye, 2006: 13).

Furthermore, some feminist writers see the potential for new digital technologies to blur the lines between humans and machines and between men and women, pointing to the use of alternative identities by people on digital media and the potential for genetic engineering, biotechnology, and cybernetics to eliminate some of the physical differences between the sexes. This demonstrates that, as well as social relations shaping technology, it is also possible for technology to shape social relations (Wajcman, 2009).

Just as it is not inevitable that new technologies will squeeze out older and less effective ones, so it is not inevitable that access to new technologies will lead to the positive social impacts initially intended by them. In both cases a whole range of factors, including affordability, social relations, motivations, and gendered and cultural norms and preferences, influence whether a technology is adopted and whether its adoption has the intended social consequences.

The case for technological momentum

While technological determinism is generally rejected by social scientists, there is a school of thought that recognizes the concept of technological momentum. In short, it is possible that investments in capital, technology, and people can create a technological system that builds sufficient speed and direction such that, as it grows, it becomes more and more difficult to change. Under such circumstances a technological system can start to have an influence and impact on its environment (Bijker, 2010). The difference between this and technological determinism is that there is human agency at the start of the process. At the outset choices are made that influence how the technology is used. It is only as momentum builds that this becomes more difficult. At a certain point in the development of the British railway system, the predominantly used distance between the two rails was agreed and became the standard gauge to allow trains to move across different companies' lines. More recently, once sufficient consumer momentum had built up behind Sony's Blu-ray™ disc format, largely due to the success of the company's PS3 console, Toshiba's rival HD DVD™ format was rendered obsolete, despite being almost identical in technical performance.

Although he didn't use the term, the economist and environmentalist E.F. Schumacher saw something close to technological momentum as a fundamental force in society (Schumacher, 1980: 42–44). In his view, some technologies are so inherently ideological that they can force society to reorganize itself so that the technology can act most efficiently, as measured by its *own* terms. An example of this would be the consequences of society's choice to adopt an industrialized form of mechanized agriculture. In countries such as the UK this led to a reduction in the agricultural labour force and a consequent decrease in the population that could be sustained in rural areas, with an associated increase in urbanization. For mechanization to operate 'efficiently', farm size has had to increase and mixed farming has been replaced by monocropping or livestock specialisms, leading to changes in the ecology of rural areas. With industrialized farming focused on competitive advantage (what can be produced most cheaply in the context of national or global markets) as opposed to what is needed for local consumption, the food distribution network has changed as we export to other parts of the country or to global markets those items we can produce most competitively and import those we no longer grow or rear locally. This, in turn, has impacted on transport infrastructure (particularly roads), the 'supermarketization' of our shops, and, ultimately, what we eat, consequently changing our health.

The adoption of the computer is another good example of technological momentum and the unpredictability of the outcome of technology choice. The primary purpose of the first computers – for example, Charles Babbage's mechanical 'difference engine' in 1822 or 'Colossus', the first electric programmable computer developed by Tommy Flowers in 1943 – was to speed up calculations (Copeland, 2008). Computers have since become the 'universal machine' or a 'general-purpose technology', applied in almost every field of human endeavour. An exponential rate of expansion of processing power has been behind this shift, driven by a doubling of the number of transistors able to fit in a dense integrated circuit every two years since the mid-1970s (Friedman, 2015). While exponential growth may pass relatively unnoticed in early years, as the doubling of something small produces something that is still relatively small, over time the impact of continual doubling becomes more pronounced and, ultimately, massive. This exponential growth has led some to suggest we are now on the edge of that massive change: a 'great restructuring' of work (Brynjolfsson and Mcafee, 2012). Whereas computers have until recently been confined to tasks that are repetitive and defined by a clear set of rules, they are now rapidly encroaching into complex communication and advanced pattern recognition, as experiments with driverless cars have shown. This application of computers in areas previously only within the capabilities of people puts whole classes of jobs at risk of being replaced by machine-based intelligence. The introduction of general-purpose technologies such as the steam engine have, in the past, also caused massive shifts in the nature of work and thus the labour market. In the case of steam, the result was rapid urbanization as populations shifted their employment from a rapidly mechanizing agriculture to factory-based production. The question raised by the computer revolution is whether, given the scale of exponential increases in processing power, the education system will be able to keep pace with changes and whether we will be able to develop new useful skills fast enough to find opportunities for all who are displaced by this advance (Brynjolfsson and Mcafee, 2012).

Modern research on science, technology, and development supports the idea that certain 'platform' technologies, such as computers, can lead society down different developmental pathways in the way Schumacher envisaged, or as the example of growing automation shows (see, for example, Smith, 2009; Leach et al., 2007). This research emphasizes our inability to predict those paths, however, and suggests that we rely more on creating wider governance arrangements that allow different parts of society to have a say in risk assessment and decision-making on investments in new technologies and applications of new science-based technologies. The call for such governance structures to be created has been seen in the areas of genetically modified material in food production, human embryo material in stem cell research, and nuclear technology for power generation (Trace, forthcoming). Managing the risk associated with technology innovation is explored in Chapter 8.

The social dimensions of technology

Social relations, social and cultural norms, preferences, and other motivations are important influences on technological innovation processes and on how technologies are disseminated and used by societies. This is an important concept as the (false) idea of technological determinism provides us the (false) sense of security that humankind will always find a technical solution to any challenge. This can have potentially catastrophic consequences. The most obvious illustration of this is our failure to mitigate climate change by reducing our material consumption patterns in the belief that other, less painful, technical solutions will eventually be found.

In reality, the invention and dissemination of new technologies can be an unpredictable and somewhat haphazard process. Genuinely new inventions, as opposed to incremental innovations on existing inventions, take time to be assimilated and for people to see their use (which may not be in line with the original intention). The phonograph was developed as a dictation machine to record important ideas and speeches; the idea of recording music came later. The telegraph was seen as a novelty and it took several years to persuade the US railroad companies of the advantage of running the lines along their rail routes. The computer was created to be an automated calculator, not the pilot of a driverless car. What is more, tools are often invented before the problem they are eventually applied to is perceived or understood: the erectile dysfunction drug marketed as Viagra® was invented as a treatment for angina, while the internet started off as a military project under the Advanced Research Projects Agency of the US Department of Defence (Mazzucato, 2013).

This is not to undermine the idea that access to technology plays a critical role in supporting human wellbeing. Neither does it contradict the fact that economic status and lack of disposable income, political power or voice in decision-making processes involving government spending often constrain a person's ability to access the technologies that are essential to a basic standard of living (issues that are explored in the rest of this book). What it does mean is that the process of developing and disseminating technologies is more complex than we might like to think and that, alongside economic status and political voice, social and cultural norms play an important role in determining the outcome.

Technological progress is 'messy' and its impacts are not always easy to predict. The choices we make around adopting new technologies can have profound consequences on society that are difficult to reverse. The concept of technological momentum shows that some technology choices can, under certain circumstances, lead to widespread environmental and societal change, with consequences far beyond those intended or foreseen.

Our inability to predict the impact of technological change poses a huge challenge for society. Understanding this 'messiness' better and thinking about how technology governance systems might respond to that challenge is what this book is about.

Note

1. An Archimedes screw is a water-lifting device consisting of a helical screw in a tube, its first record of use being around 250 BC in Egypt.

References

Baron, E. and Magone, C. (2014) 'The polio eradication campaign: time to shift the goal', *International Health* <http://dx.doi.org/10.1093/inthealth/ihu004>.

Begum, S. (1983) 'Women and technology: rice processing in Bangladesh', *Women in Rice Farming: Proceedings of a Conference on Women in Rice Farming Systems,* pp. 221–41, Manila: International Rice Research Institute.

Bijker, W.E. (2010) 'How is technology made? – That is the question!', *Cambridge Journal of Economics,* 34: 63–76 <http://dx.doi.org/10.1093/cje/bep068>.

Brynjolfsson, E. and Mcafee, A. (2012) *Race Against the Machine: How the Digital Revolution is Accelerating Innovation, Driving Productivity, and Irreversibly Transforming Employment and The Economy,* Cambridge, MA: MIT Sloan School of Management.

Buchanan, R.A. (2014) *History of technology* [online]. Encyclopedia Britannica <http://www.britannica.com/EBchecked/topic/1350805/history-of-technology> [accessed 17 March 2015].

Channel 4 (2014) '#Jihad: how Isis is using social media to win support', 17 June, <www.channel4.com/news/isis-iraq-social-media-jihad-billion-campaign-recruit-video> [accessed 11 March 2016].

Compton, A.H. (1940) *The Human Meaning of Science,* p. 2, Chapel Hill, NC: University of North Carolina Press.

Copeland, B. (2008) *The modern history of computing* [online]. Stanford Encyclopedia of Philosophy <http://plato.stanford.edu/archives/fall2008/entries/computing-history/> [accessed 11 January 2016].

Cross, J. (2013) 'The 100th object: solar lighting technology and humanitarian goods', *Journal of Material Culture,* 18: 367–87 <http://dx.doi.org/10.1177/1359183513498959>.

d.light (2014, 22 July) *d.light reaches 500,000th pay-as-you-go customer, launches access initiative* [online]. d.light <http://www.dlight.com/files/1514/1739/2300/7-22-14_d.light_500K_Financed__Access_Accelerator.pdf> [accessed 27 March 2015].

Friedman, T.L. (2015) 'Moore's law turns 50', *The New York Times,* 13 May <http://www.nytimes.com/2015/05/13/opinion/thomas-friedman-moores-law-turns-50.html?_r=0>.

Harper, D. (2015). *Technology.* Online Etymology Dictionary <http://dictionary.reference.com/browse/technology> [17 March 2015].

Howard, P., Duffy, A., Freelon, D., Hussain, M., Mari, W., and Mazaid, M. (2011) *Opening Closed Regimes: What was the Role of Social Media during the Arab Spring?* Seattle, WA: University of Washington.

Leach, M., Scoones, I., and Stirling, A. (2007) *Pathways to Sustainability: An Overview of the STEPS Centre Approach Paper.* Brighton: University of Sussex, STEPS Centre.

Mazzucato, M. (2013). *The Entrepreneurial State: Debunking Public vs. Private Sector Myths*. London: Anthem Press.

Nye, D.E. (2006). *Technology Matters. Questions to Live With*. Cambridge, MA: The MIT Press.

Schumacher, E. (1979) *Good Work*, New York: Harper and Row.

Smith, J. (2009) *Science and Technology for Development*. London: Zed Books.

Stilgoe, J., Owen, T., and Macnaghten, P. (2013) 'Developing a framework for responsible innovation', *Research Policy*, 14: 1568–80 <http://dx.doi.org/10.1016/j.respol.2013.05.008>.

Trace, S. (forthcoming) 'Technology and development', in P.H. Callan, *The International Encyclopaedia of Anthropology*, Hoboken, NJ: Wiley-Blackwell.

SE4All. (2013). *Global Tracking Framework*. New York: United Nations Sustainable Energy for All Initiative.

Wajcman, J. (2009) 'Feminist theories of technology', *Cambridge Journal of Economics* <http://dx.doi.org/10.1093/cje/ben057>.

CHAPTER 2

Technology Justice: establishing the principle

> *Science and technology have freed humanity from many burdens and given us this new perspective and great power. This power can be used for the good of all. If wisdom governs our actions; but if the world is mad or foolish, it can destroy itself just when great advances and triumphs are almost without its grasp.*
>
> Jawaharlal (Pandit) Nehru (quoted in Vinod and Deshpande, 2013: 507)

The prologue and introduction to this book described environmental sustainability and ending poverty as the two greatest challenges facing humankind today. They also set out some initial examples to demonstrate the importance of technology to both challenges and suggested that the absence of a system of governance for technology is a major obstacle to addressing either. This lack of governance at a critical time in human history is the cause of great injustice, to present and future generations, something that will be discussed further in later chapters. A new concept of Technology Justice was called for to act as the focus for a series of radical changes required to 'reboot' our relationship with technology. The purpose of this chapter is to explore what such a concept might look like by first reviewing the links between technology and the environment, and technology and human wellbeing in more depth, and then considering the notion of a safe, just, and inclusive space for humanity to operate in, before finally proposing a definition for a Technology Justice principle.

Technology and environment

In *Small is Beautiful*, E.F. Schumacher was one of the first voices of the then emerging environmental movement to argue that humankind was facing a consumption crisis. He pointed out that if the quantities of materials consumed by Europe and North America were replicated globally, the world's natural resources would be insufficient to meet this demand, a point echoed 40 years later by the World Wildlife Fund which estimated that four planets worth of resources would be needed to sustain everyone on the planet today at US consumption levels (WWF, 2012). Schumacher's case for treating natural resources as 'capital' rather than recurrent income, to be drawn down only with great care and as part of a process of moving to a more sustainable resource base, has since been echoed in other landmark publications, including *The Limits to Growth*[1] (Meadows et al., 1972) and the Brundtland Report, *Our Common Future* (The World Commission on Environment and Development, 1987).

In 2009, a group of environmental scientists and earth system academics led by Johan Rockström of the Stockholm Resilience Centre and Will Steffen

http://dx.doi.org/10.3362/9781780449043.003

of the Australian National University proposed a framework of 'planetary boundaries'. The planetary boundary approach raises the question: 'What are the non-negotiable planetary preconditions that humanity needs to respect in order to avoid the risk of deleterious or even catastrophic environmental change at continental to global scales?' (Rockström et al., 2009). The approach goes beyond considering environmental limits simply in terms of the finiteness of resources to thinking about the impact of the use of those resources on the functioning of ecological systems or earth processes critical to human wellbeing.

Nine broad earth processes have been identified by Rockström's team, as shown in Figure 2.1, with two (biosphere integrity and biogeochemical flows) being each further divided into two subcategories. The idea behind positing boundaries for these processes is that tipping points or values exist at which very small further increases produce a large and possibly irreversible, catastrophic change: the release of CO_2 into the atmosphere leading to global warming that triggers a collapse of the polar ice sheets, or ocean acidification causing a collapse of the marine food chain, for example. As the earth's system is very complex and these variables do not exist in isolation from each other, the exact location of tipping points is difficult to predict. The planetary boundary approach therefore establishes a range of possible values within which the tipping point is thought to lie for each process, with the lower end of that range defined as the edge of the safe space and the beginning of a zone of uncertainty and danger for humankind. Using this approach, Rockström's team has assessed that four of the processes (climate change, the biogeochemical flows, land-use change, and rate of biodiversity loss) have already breached the safe limits proposed, while the boundaries for two others (novel entities and atmospheric aerosol loading) have yet to be established.

The concept and choice of planetary boundaries has had its critics.[2] The Breakthrough Institute, for example, agreed that real biophysical threshold elements exist in the global climate system and that there are global thresholds for ocean acidification, ozone depletion, and phosphorus levels, but argued that only local tipping points exist for the remaining 'boundaries', suggesting that land-use change, freshwater use, or nitrogen levels in one region, for instance, are ecologically independent of these processes and their impacts in other regions (Nordhaus et al., 2012). Meanwhile, some ecologists have argued that the idea of safe boundaries is itself unhelpful as, instead of trying to nip potentially dangerous activity in the bud, it encourages harmful human activity right up to the limit, even for processes such as freshwater use and ozone depletion that, for the moment at least, are some way from their respective boundaries (see, for example, Schlesinger, 2009).

Nevertheless, the concept of planetary boundaries has been widely adopted by groups ranging from non-governmental organizations, including the World Wildlife Fund (WWF, 2011) and Oxfam (Raworth, 2012), to international bodies, such as the United Nations (for example, see UN, 2012), as a useful tool to inform governance and policy discussions on ways to achieve environmental sustainability. It is adopted in this book for the same reason.

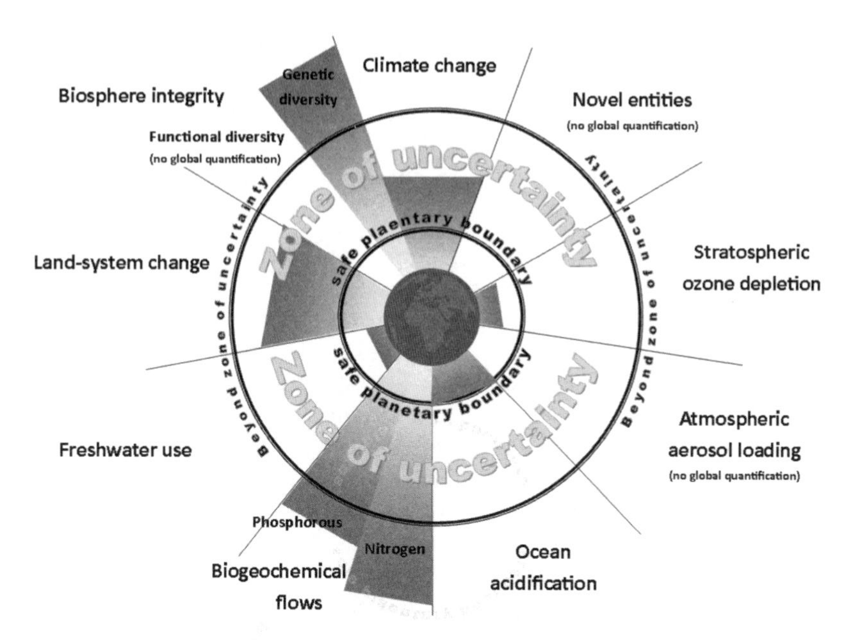

Figure 2.1 Rockström's safe planetary boundaries
Source: Stockholm Resilience Centre, 2015 (credit F. Pharand-Deschênes/Globaïa)

A critical question posed by this analytical framework is 'What is it we need to do to reduce the human-induced breaches of the safe planetary boundaries where they have already occurred and to prevent other boundaries from being breached in the future?' Part of the answer to that question has to be non-technical: we need to reduce our levels of material consumption to fit within the carrying capacity of the planet. Such changes in consumption will require changes in attitude to almost every aspect of our lives, including some less obvious ones: for example, eating less meat to reduce land clearance to grow ever more feed for livestock and to reduce the rising methane emissions from livestock themselves. This is a huge behavioural change project with massive political and economic implications.

Alongside behavioural change, there are major technological challenges to overcome for every one of the nine planetary boundaries identified. Many of these challenges are already well known and discussed in policy debates. In the case of controlling climate change, the massive technological changes required across a range of sectors include the following:

- In energy production, technical improvements in energy efficiency combined with increased use of renewable energy technologies will be needed to reduce CO_2 emissions (SE4All, 2013).
- In agriculture, given the scale of greenhouse gas emissions (methane and nitrous oxide as well as CO_2) arising from the production and application of inorganic fertilizers and pesticides, a shift to more ecological and

low-input forms of farming is required (see IAASTD, 2009; Camargo et al., 2013).

- In industry, substitutes for high-energy processes that emit large amounts of CO_2 are needed, such as an alternative to the calcination of limestone for cement, a process that is responsible for 5 per cent of all greenhouse gas emissions (Rubenstein, 2012).

Many areas of necessary technological innovation have the potential to touch multiple earth system processes and so address multiple tipping points in the planetary boundary model. Improving energy efficiency and shifting to renewable energy generation will impact not only climate change through reduced greenhouse gas emissions, but also slow down ocean acidification, a process which is fed by increased concentrations of CO_2 in the atmosphere (Fabry et al., 2008). It will also reduce freshwater use, as fossil fuel power production accounts for around 15 per cent of global freshwater extraction (IEA, 2012). Likewise, a reduction in the use of chemical fertilizers and pesticides will not only cut the greenhouse gas emissions arising from their production and application, but also reduce the burden on terrestrial and marine ecologies resulting from overuse of nitrogen and phosphorus in agriculture, and diminish the negative impacts of pesticides on biodiversity (IAASTD, 2009).

Table 2.1 shows the planetary boundaries and the expected consequences of crossing their respective tipping points. It also contains examples of the technological innovation and change required to address these threats.
Table 2.1 also illustrates how technology is both at the core of the threat of planetary boundary breach (for example, greenhouse gas emissions from the use of fossil fuel technologies) and a potential part of a solution (to use the same example, through innovation in renewable energy technologies and improvements in energy efficiency). It also shows the scale of the consequence, and thus the scale of the injustice that would be visited on future generations, if we fail to find an effective system to govern technology use and innovation. Technology Justice, in this respect, must mean establishing a governance system that can manage and reduce the very real and substantial environmental risks that stem from today's use of technology, while simultaneously steering innovation processes to deliver the new green technologies needed to keep humankind safely within the planetary boundaries.

Technology and human wellbeing

Much research time and academic effort has gone into trying to define and measure human wellbeing. Martha Nussbaum used Amartya Sen's capability approach (discussed in detail later in this chapter) to articulate a conception of wellbeing based on 10 fairly abstract capabilities, including the ability to live a life of normal length, to have bodily health and bodily integrity, to be able to use the senses to imagine and think, to be able to relate to others and have a basis for self-respect, to be able to express and relate to emotion, to exercise

Table 2.1 Expected consequences of crossing planetary boundaries and the technological challenges of staying within them

Planetary boundary	*Expected consequences of crossing planetary boundaries*	*Examples of technological changes required to stay within the boundaries*
Climate change	Release of greenhouse gases leads to global temperature rise; loss of polar ice sheets and glacial freshwater supplies; rapid sea-level rise; bleaching and mortality in coral reefs; more frequent large floods; abrupt shifts in forest and agricultural systems, all potentially challenging the viability of human societies	*Energy*: shift from coal, oil, and gas to modern clean renewable energy sources; radical improvements in energy efficiency; exploration of viability of carbon capture and storage technologies *Agriculture and food*: major reduction in fertilizer use by shifting to low-input agroecological technologies for human food and animal feed production *Industrial processes*: find substitutes for energy-intensive processes, for example an alternative to the calcination of limestone for cement
Biosphere integrity	Reduced resilience of land and marine ecosystems, especially in the face of climate change and increasing ocean acidity; large-scale biodiversity loss, possibly leading to sudden and irreversible consequences for ecosystems	*Energy*: shift to renewables and improvements in energy efficiency to avoid burning fossil fuels and biomass *Agriculture and food*: major reduction in pesticide and herbicide use by shift to agroecological technologies; shift from monoculture to more mixed-farm ecologies; conservation of biodiversity in food chain (livestock breeds and crop varieties); scientific management of fish stocks / control of fishing technologies *Industrial processes*: recycling technologies to reduce demand for raw materials *Urban environment*: improvements to urban planning, for example wastewater and solid waste management, green space, efficient transport systems
Nitrogen use	Raised acidity of soils, and algal blooms in coastal and freshwater systems that deplete oxygen levels, pollute waterways, and kill aquatic life thus threatening the quality of air, soil, and water, and eroding the resilience of other earth systems	*Energy*: shift to renewables and improvements in energy efficiency to avoid burning fossil fuels and biomass[1] *Agricultural and food*: major reduction in fertilizer use by shift to low-input agroecological technologies for human food and animal feed production; improvements in capture and treatment of livestock manure *Urban environment*: improvements in the capture and treatment of human waste

(*continued*)

Table 2.1 Expected consequences of crossing planetary boundaries and the technological challenges of staying within them (*continued*)

Planetary boundary	*Expected consequences of crossing planetary boundaries*	*Examples of technological changes required to stay within the boundaries*
Phosphorus use	Depleted oxygen levels in freshwater bodies and coastal waters risking abrupt shifts in lake and marine ecosystems	*Agriculture and food*: major reduction in fertilizer and pesticide use by shift to low-input agroecological technologies for human food and animal feed production *Industrial processes*: alternative to sodium tripolyphosphate for use as detergent builder *Urban environment*: implementation of more efficient urban wastewater treatment processes
Freshwater use	Shifts in regional rainfall and climate (for example, the monsoon), and reduced biomass production and biodiversity, decreasing the resilience of land and marine ecosystems, and undermining human water supply, food security, and health	*Energy*: reduction in water abstraction for cooling and steam in power stations through shift to renewables and improvements in energy efficiency *Agriculture and food*: ultra-efficient irrigation technologies, rainwater harvesting, techniques to improve moisture retention in soil, and breeding of low water-requirement crops *Industrial processes*: improved efficiency and reuse to reduce water abstraction for industry; efficient renewable energy-based processes for desalinizing sea water *Urban environment*: rainwater harvesting and grey water recycling
Land-use change	Serious threat to biodiversity and to the regulatory capacities of the earth system by affecting the climate system and the freshwater cycle	*Energy*: alternative cooking fuels to reduce burden on forests from fuelwood demands; limits on use of biofuels from plant materials *Agriculture and food*: shift to agroecological food production processes more aligned to the natural environment; improvements in storage to reduce losses and therefore land requirements *Urban environment*: reduce environmental footprint of urban centres through local production of renewable energy; improvements in energy and water-use efficiency; recycling, waste management, and urban agriculture

(*continued*)

Table 2.1 Expected consequences of crossing planetary boundaries and the technological challenges of staying within them (*continued*)

Planetary boundary	*Expected consequences of crossing planetary boundaries*	*Examples of technological changes required to stay within the boundaries*
Ocean acidification	Increased CO_2 being dissolved in sea water leading to loss of calcifying marine organisms; serious impacts on the productivity of coral reefs with likely ripple effects up the food chain	*Energy*: as per climate change *Agriculture and food*: as per climate change *Industrial processes*: as per climate change
Stratospheric ozone depletion	Severe and irreversible ultra-violet radiation with especially damaging effects on marine ecosystems, and on the health of humans exposed to radiation	*Industrial processes*: alternatives to the use of chlorofluorocarbons in refrigerators, airconditioners, and aerosol cans
Atmospheric aerosol pollution	Release of fine particles into the air, primarily through burning fossil fuels and biomass, leading to: changing global rainfall patterns, including monsoon systems; damage to crops and forests; acid rain killing fish; human health impacts and premature death due to respiratory disease	*Energy*: as per climate change *Agriculture and food*: as per climate change *Industrial processes*: as per climate change
Release of novel entities (synthetic substances and novel life forms)	Reduced abundance of species, likely to create bioaccumulation of effects up food chains, with impacts on human immune systems and neurodevelopment; likely to increase vulnerability of organisms to stresses such as climate change	*Industrial processes*: increase focus on green chemistry; increase learning from previous mistakes; and improved application of the precautionary principle to the use and release of novel entities[2]

1. The burning of fossil fuels releases nitrogen oxide (in addition to CO_2) which, in turn, can be washed out of the atmosphere onto land and water bodies, further increasing the nitrogen burden on the marine and terrestrial environment (EPA, 2015).

2. For more on green chemistry see Steffen et al. (2015), Sanderson (2011), and Schulte et al. (2013). For more on the precautionary principle and the release of novel entities see Chapter 8 of this book and European Environment Agency (2013).[/N]

Source: The 'expected consequences' column of this table draws directly on material presented by Kate Raworth (2012: 17), which in turn draws on the original paper on planetary boundaries published by Rockström et al. (2009)

practical reason, to play, and to be able to exercise control of one's political and material environment (Nussbaum, 2003). A more recent seven-year programme (2002–09) of cross-cultural research at the University of Bath aimed to develop a definition and means of measuring wellbeing more amenable to use by development practitioners. The work included a study into local definitions of wellbeing in Bangladesh, Ethiopia, Peru, and Thailand (Camfield, 2006) and concluded that there were three common components of local constructions of the concept that transcended cultural differences (White, 2009):

- a material aspect: that a person's basic needs – food, shelter, access to basic services such as water and energy, education and health, and an income to pay for all of this – are met;
- a relational aspect: that a person has a degree of control over their own life; they can be a part of decisions that have a major impact on the way they live; they can live in dignity with the respect of their fellow citizens; and they can live in peace with their neighbours;
- a subjective aspect: that a sense of wellbeing is also influenced by people's perception (and not just the objective reality) of the material and relational aspects.

More recently still, an economist working for Oxfam, Kate Raworth, was inspired by Rockström and colleagues' planetary boundaries to develop a similar framework proposing a set of social boundaries. Raworth used a review of national and regional government submissions on social priorities at the Rio+20 conference to develop a framework with 11 elements (which largely fit into the material and relational elements of White's analysis) to describe what she termed a minimum 'social foundation' (Raworth, 2012). These 11 elements are listed in Table 2.2, along with Raworth's description of each and her analysis of what proportion of the world's population has failed to achieve those minimum standards of living.

Using Rockström's presentation of planetary boundaries as a guide, Raworth represents this information about progress towards universal achievement of a social foundation, as shown in Figure 2.2.

What is striking from Table 2.2 and Figure 2.2 is that of the eight elements of Raworth's social foundation against which progress can be assessed with currently available data, not a single one has yet been extended universally. Of the global population, 13 per cent remains malnourished, 21 per cent have yet to reach a minimum income of $1.25 per day, 30 per cent lack access to the World Health Organization's list of essential medicines, and so on.

It can be argued that Raworth's list of 11 elements is a rather arbitrary definition of a social foundation, based as it is on a single source of suggestions. Certainly, the exclusion of adequate shelter as part of a comprehensive social foundation is an omission, while the categories of 'jobs', 'income', and 'social equity' all seem to overlap. That said, adjusting it to reflect the final version of the social elements of the Sustainable Development Goals, as the internationally accepted set of development targets, would not result in huge change. For the purpose of the arguments made here, Raworth's elements

Table 2.2 Elements of a social foundation and proportion of global population below this floor

Social foundation	*Extent of global deprivation*	*Percentage below floor (%)*[1]	*Year of data*
Food security	Population undernourished	13	2006-08
Income	Population living below US$1.25 (PPP) per day	21	2005
Water and sanitation	Population without access to an improved drinking water source	13	2008
	Population without access to improved sanitation	39	2008
Health care	Population estimated to be without regular access to essential medicines	30	2004
Education	Children not enrolled in primary school	10	2009
	Illiteracy among 15–24 year olds	11	2009
Energy	Population lacking access to electricity	19	2009
	Population lacking access to clean cooking facilities	39	2009
Gender equality	Employment gap between women and men in waged work (excluding agriculture)	34	2009
	Representation gap between women and men in national parliaments	77	2011
Social equity	Population living on less than the median income in countries with a Gini coefficient[2] exceeding 0.35	33	1995–2009
Voice	For example, population living in countries perceived (in surveys) not to permit political participation or freedom of expression	To be determined	
Jobs	For example, labour force not employed in decent work	To be determined	
Resilience	For example, population facing multiple dimensions of poverty	To be determined	

Notes: PPP = purchasing power parity

1. Compiled from publicly available statistics from institutions such as the World Bank and the UN.

2. The Gini coefficient is a statistical measure of inequality, commonly applied to income, where 0 equals absolute equality, where all values are the same, and 1 equals maximum inequality, for example where just one person has all the income.

Source: Raworth, 2012: 10

suffice. Together they demonstrate that it is possible to define a minimum social foundation, largely using existing internationally accepted norms. Moreover, using those available standards, it is obvious that the world is still a long way from establishing such a universal social foundation.

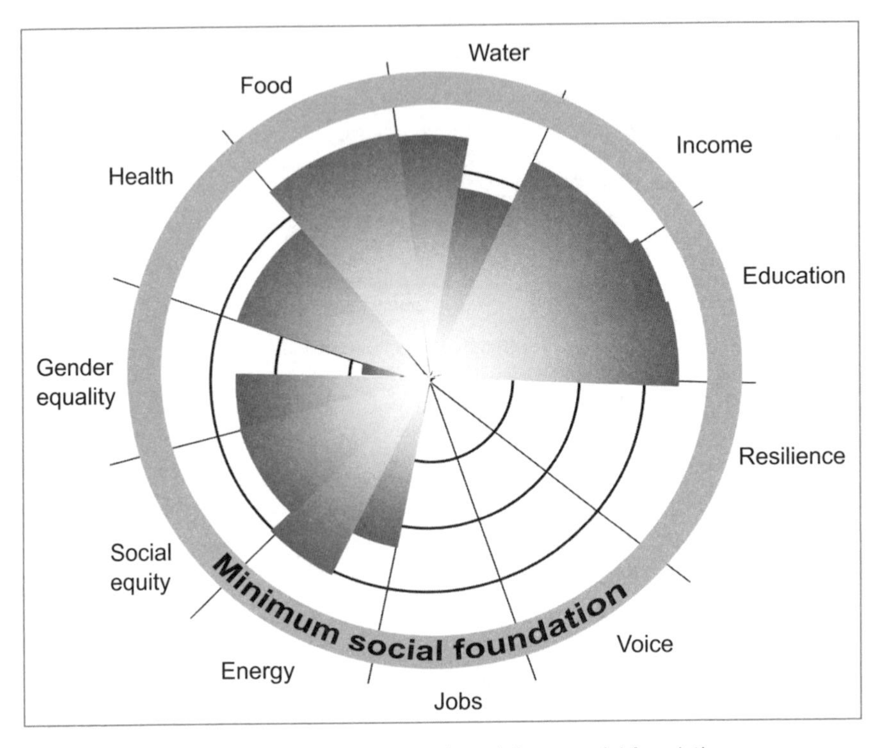

Figure 2.2 Proportion of global population meeting minimum social foundation
Source: Raworth, 2012

Technology has an important and often critical role to play in closing the remaining gaps where people fall below the social foundation. For example, access to energy technologies (electricity, fuels, and motive power) is a fundamental enabler without which many activities cannot take place. Energy technologies are critical for cooking, lighting, heating, cooling, and communications in the home, but also to support the provision of community services such as security (street lighting), health (to light clinics, refrigerate vaccines, and power diagnostic equipment), education (to light study areas, power computers and the internet, and enable cooking of school meals), and the creation of livelihoods in agriculture (for example, to power irrigation) or in small businesses and industry (to drive equipment and plant) (Practical Action, 2014). As such, they touch directly or indirectly on all 11 elements of Raworth's foundation. Water-supply technology similarly acts as a cross-cutting enabler, supporting everything from food production to health service provision. It is also a technology that has huge implications for gender equity, impacting as it does on the burden of water collection, a domestic role that is largely performed by women and girls (Jansz and Wilbur, 2013). Access to technical knowledge can also be an important part of improving living standards across many of the elements of the social foundation, being vital to improving productivity in agriculture and other forms of enterprise, or improving the diagnosis and treatment of disease, as examples.

Technology is obviously not the only condition to be fulfilled to achieve each element of the social foundation. Food security is not dependent solely on the technical capability to produce, transport, and store food. It is also controlled by conditions of entitlement: the ability to pay for food or be eligible for other means of support to access food in times of difficulty. Likewise, voice and social equity clearly depend largely on the establishment of political freedoms and gender equality in challenging culturally ascribed gender roles. But the absence of technology is almost always a key feature of living in extreme poverty and, as the further examples in Table 2.3 show, access to technology either contributes to or is a necessary condition for the achievement of all elements of the social foundation.

Table 2.3 Technologies required to support Raworth's social foundation

Social foundation element	*Examples of access to technology required to eliminate deprivation*
Food security	*Agricultural*: improved seeds and livestock, irrigation, agro-processing, and storage technologies
	Energy: for irrigation, cultivation, food processing, and preservation of food on and off farm
	Water: for irrigation
	Transport: distribution of inputs and products
	Technical knowledge: to improve productivity
Income	*Agricultural*: as per food security to improve viability of rural livelihoods
	Energy: to power rural and urban enterprises
	Water and sanitation: water supply for enterprise processes
	Industrial: machine tools and equipment for small, medium, and large-scale industry
	Communications: market information and marketing, maintaining supply chains
	Technical knowledge: technical skills and knowledge for rural or urban enterprises
	Transport: for goods and for access to work
Water and sanitation	*Water and sanitation*: water extraction/harvesting and delivery technologies; sanitation technologies
	Energy: for water pumping, desalinization of water for drinking, and treatment processes for drinking water and sewage
	Technical knowledge: for design operation and maintenance of systems
Health care	*Medical technologies*: medicines, medical equipment
	Energy: to keep health facilities open after dark, to operate medical equipment, to maintain cold chain for vaccines, to power communications
	Transport: for patients' access and for medical supplies
	Water and sanitation: maintenance of hygiene in health facilities
	Communications: remote diagnosis, maintaining supply chains
	Technical knowledge: medical skills

(*continued*)

Table 2.3 Technologies required to support Raworth's social foundation (*continued*)

Social foundation element	*Examples of access to technology required to eliminate deprivation*
Education	*Energy*: for lighting and heating/cooling in classrooms, for water supply, preparation of school meals, to enable communications
	Water and sanitation: technologies for students and teachers
	Communications: information and communication technologies, access to internet and learning resources
Energy	*Energy*: electricity (off-grid or on-grid), clean cook-stoves, and motive power
	Technical knowledge: for design, operation, and maintenance of systems
Gender equality	*Energy*: technologies can reduce disproportionate time women and girls spend on gender-ascribed roles related to fuel collection and cooking, impacting on school attendance and time for livelihood activities
	Water and sanitation: freeing up time for women and girls, as per energy; sanitation facilities for girls at school increases attendance
	Technical knowledge: access for women and girls to provide new employment opportunities and to challenge gender stereotyping of roles
Social equity	Access to technologies as per Income element to improve distribution of employment opportunities and thus income distribution, contributing to improvement in Gini coefficient
Voice	*Communication*: access to information; ability to communicate and organize
	Transport: freedom of movement; ability to communicate and organize
Jobs	As per Income
Resilience	See all the above (technology supporting reduction in multiple dimensions of poverty)

Access to a basic suite of technologies is essential to establishing a basic standard of living or social foundation for everyone on this planet. The technologies needed to achieve this, as can be seen from Table 2.3, already exist and indeed are already in use and enjoyed by the majority of the world's population. But a substantial minority, as shown in Table 2.2, have yet to achieve the social foundation. This is not merely unfortunate. It is an immense injustice that humanity has not managed to ensure universal access to technologies critical to achieving a minimum reasonable standard of living, technologies that have generally been in existence and use for decades and, in some cases, centuries. Technology Justice, in this respect, must mean establishing a global governance process that ensures these gaps in technology access are addressed and closed, something it has long been in our power to do.

A safe operating space for humanity

The two great challenges we face are ending poverty while achieving environmental sustainability. The analysis above shows that humankind has already broken through several elements of a safe environmental ceiling or planetary boundary. Rockström's team talks about 'an urgent need for a new paradigm that integrates the continued development of human societies and the maintenance of the Earth System in a resilient and accommodating state' (Steffen et al., 2015: 736). They also suggest that the planetary boundaries framework defines a safe operating space for humanity. But Raworth's argument is that the planetary boundaries only delineate one aspect of that safe operating space. Deep and even existential threats are also posed to the 40 per cent of the world's population living on less than $2 per day, most of whom will be exposed to significant risk from a failure to reach the thresholds of many elements of a basic social foundation. Raworth says we need to go beyond the idea of an environmentally safe operating space and aim to achieve a safe, inclusive, and sustainable space for human development. Combining the planetary boundary and social foundation models, as shown in Figure 2.3, she suggests we need a new form of 'doughnut economics' (named after the shape of the resulting diagram) that focuses on moving humankind into that safe annular space between a social foundation and planetary boundaries (Raworth, 2012).

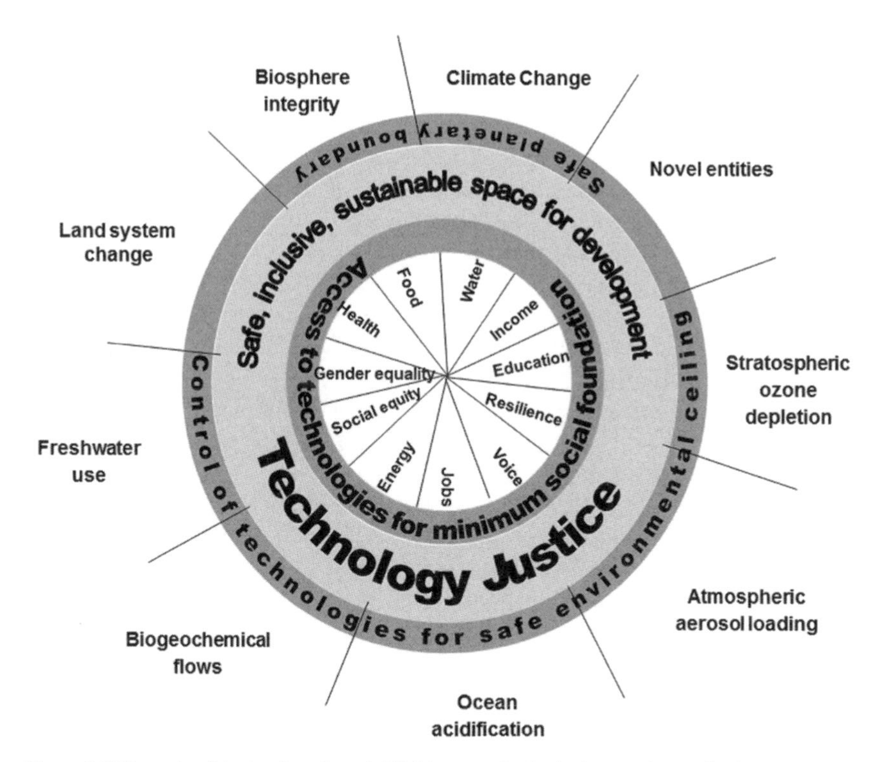

Figure 2.3 The role of technology in establishing a safe, inclusive, and sustainable space for development

Source: Raworth, 2012

Access to technology is critical to establishing a social foundation. But better control of technology use, combined with further rapid innovation in green technologies, will be critical to remaining within planetary boundaries. The boundary conditions for technology to be used in a just manner thus map directly onto Raworth's boundaries for a safe, inclusive, and sustainable space for development and form an important part of the new paradigm of 'doughnut economics'. Raworth's safe space for humanity thus also represents Technology Justice, a space where technology is used to help achieve a minimum reasonable standard of living for everyone while staying within the ecological carrying capacity of the planet.

Defining a principle of Technology Justice

So how should a principle that could be used to guide a paradigm shift in the governance of technology be formulated? Firstly, bearing in mind that technological progress is not inevitable and that the social impact of technology is not always predictable, defining Technology Justice in terms of a broad principle that can be used in a reflexive and iterative manner is likely to be more useful than a rigid set of rules to be applied, for example, at the outset of a technology development process. But what model of justice should this principle appeal to or adopt? There is much to choose from but two relatively recent constructs are briefly related here to establish the principle of Technology Justice: one from a societal standpoint and one from an individualistic standpoint.

In his 1971 book *A Theory of Justice*, the American philosopher John Rawls conceptualized justice as fairness. He suggested that the principles of justice should be decided from behind what he called a 'veil of ignorance'. The veil of ignorance essentially enables someone to frame the rules and organization of the society they are about to enter, but know nothing about themselves (their material wealth, intelligence, physical strength, social standing, and so on) and therefore nothing about the place they will take up in that society. In this position it is difficult for that person to tailor the principles of justice in a way that would advantage themselves over others as they lack the necessary information. Essentially, if someone had to develop principles of justice from behind the veil of ignorance, they would have to develop principles that treated everyone fairly and, in particular, maximized the prospects of the least well-off.

From a position of justice as fairness we can examine whether the current distribution of opportunity to access certain technologies is just or unjust. Assuming that there are a certain set of technologies that are widely available and essential to establishing and maintaining a basic minimum standard of life or wellbeing, then a person considering the distribution of those technologies from behind a veil of ignorance would want to ensure, as a principle of justice, that those most disadvantaged (in case that is the position in society they eventually occupy) have the opportunity to access those technologies on a par with the rest of society.

It is also possible to use the idea of justice as fairness to examine one person's use of technology and its resulting impact on another person's life. Taking greenhouse gas emissions from fossil fuel technology as an example, from behind a veil of ignorance and not knowing whether one will be born today or in the future, as a matter of justice one would wish for technologies to be used in a way that avoided negative impacts not only for today but also in the future. The idea that justice might require trade-offs or negotiation still sits well with Rawls's view of justice as fairness. In particular, it aligns with his first principle of justice, namely that: 'each person is to have an equal right to the most extensive scheme of equal basic liberties compatible with a similar scheme of liberties for others' (Rawls, 1999: 53). Justice as fairness, both fairness in terms of opportunity to access a technology and fairness in terms of the impact the use of that technology has on others, requires the application of compromise, both within and between generations.

Amartya Sen, Nobel Prize-winning economist and pupil of Rawls, provides a more individualistic approach to justice in defining a standard of living. Sen's work on defining the components of a standard of living has been one of the most influential attempts to introduce the concept of wellbeing into definitions of development and was instrumental in the creation of the UN Human Development Index. Sen rejected rising interest in the use of happiness[3] to define an acceptable standard of living on the basis that social conditioning can mean that even a very deprived person can still express happiness. He concluded that it is morally wrong to label happiness in such circumstances as an indicator of wellbeing. Instead, he argued, a person's wellbeing in a society depends on them being capable of carrying out certain key functions (for example, feeding themselves, being healthy, having a good job, being safe, being able to appear in public without shame). Functions are the various things a person may value doing or being and it is the capability to carry out these functions in the context of a particular society that defines wellbeing, and a reasonable standard of living, in that society (Sen, 1985). For Sen, the notion of freedom of choice is also critical to a definition of wellbeing, which is why he focuses on a person's capacity to carry out key functions as opposed to whether the functions are actually performed. He uses the example of a starving child and a fasting monk to illustrate the point. Both are failing to perform the function of adequately feeding themselves. But the fasting monk has the capability to fulfil the function and also the freedom to choose not to. He has the ability to live the life he values. The starving child has no choice and cannot live the life she values (Sen, 1985).

For Sen, justice is the freedom for people to live the life they value. Technology fits well into this framework as a means to enable people's capabilities to carry out the functions necessary for this freedom (see, for example, Hatakka and De', 2011; Oosterlaken, 2013). However, echoing the discussion earlier, Sen's capabilities approach, while hugely influential, has been criticized for focusing too much on notions of individual freedom and failing to recognize that one person's freedom to live the life they value may well compromise another person's freedom to do the same (Deneulin and McGregor, 2009). The UK government's Sustainability Commission has suggested the alternative notion of capabilities to flourish being bound by ecological limits (Jackson, 2009), while the researcher and writer Robert Chambers has similarly offered the idea of 'responsible wellbeing' (Blackmore, 2009) to cope with this criticism.

It is this sense of Sen's capabilities approach tempered by Rawls's theory of justice as fairness and compromise that the principle of Technology Justice seeks to capture. Everybody should have the right to the technologies needed to live the life they value, tempered by the recognition that ecological limits and notions of fairness mean we must avoid diminishing the ability of others, now and in the future, to realize the same right.

To be effective, such a principle should apply not only to ensuring just access to and use of existing technologies. Given the existential crises of poverty and environmental catastrophe the world faces today, a principle of Technology Justice would also need to guide future technological innovation efforts so that they, too, allow subsequent generations everywhere to live a life they value.

Technology Justice is therefore a principle or vision of a world where:

- Everyone has access to the technologies needed to achieve a reasonable standard of living in a way that doesn't prevent others now and in the future from doing the same.
- The focus of efforts to innovate and develop new technologies is firmly centred on solving the great challenges the world faces today – ending poverty and providing a sustainable future for everyone on our planet.

It is this principle that underpins the analysis and arguments in the following chapters.

Notes

1. The book *Limits to Growth*, published in the same year as *Small is Beautiful* and the cause of much debate, used a computer simulation of exponential economic and population growth with finite resource supplies. It explored three future scenarios, two of which saw 'overshoot and collapse' of global systems by the second half of the 21st century, while the third resulted in a 'stabilized world' (Meadows et al., 1972).
2. For a rebuttal of these and other critiques see Rockström (2012).
3. For an example of the use of happiness in defining and measuring well-being, see the Government of Bhutan's Gross National Happiness Index (Centre for Bhutan Studies and GNH Research, 2015).

References

Blackmore, C. (2009) *Responsible Wellbeing and its Implications for Development Policy*, University of Bath Wellbeing in Developing Countries Working Paper 09/47, Bath: University of Bath.

Camargo, G., Ryan, M., and Richard, T. (2013) 'Energy use and greenhouse gas emissions from crop production using the farm energy analysis tool', *BioScience* 63: 263–73.

Camfield, L. (2006) *The Why and How of Understanding 'Subjective' Wellbeing: Exploratory Work by the WeD Group in Four Developing Countries*, WeD Working Paper 26, Bath: ESRC Research Group on Wellbeing in Developing Countries.

Centre for Bhutan Studies and GNH Research (2015) Bhutan's 2015 Gross National Happiness Index [online] <http://www.grossnationalhappiness.com/> [accessed 15 January 2015].

Deneulin, S. and McGregor, J. (2009) *The Capability Approach and the Politics of a Social Conception of Wellbeing*, University of Bath Wellbeing in Developing Countries Working Paper 09/43, Bath: Bath University.

EPA (2015) *Nutrient pollution – the sources and solutions: fossil fuels* [online] United States Environment Protection Agency <http://www.epa.gov/nutrientpollution/sources-and-solutions-fossil-fuels> [accessed 13 January 2016].

European Environment Agency (2013) *Late Lessons from Early Warnings: Science, Precaution, Innovation*, Copenhagen: European Environment Agency.

Fabry, V., Seibel, B., Feely, R., and Orr, J. (2008) 'Impacts of ocean acidification on marine fauna and ecosystem processes', *ICES Journal of Marine Science* 65: 414–32.

Hatakka, M., and De', R. (2011) 'Development, capabilities and technology – an evaluative framework', in *Proceedings of the 11th International Conference on Social Implications of Computers in Developing Countries, Kathmandu, Nepal, 22–25 May 2011*. Kathmandu: International Federation for Information Processing <http://www.ifipwg94.org/publications> [accessed 4 March 2016].

IAASTD (2009) *Agriculture at a Crossroads: IAASTD Synthesis Report*. Washington: International Assessment of Agricultural Knowledge, Science and Technology for Development.

IEA (2012) *Water for Energy: Is Energy Becoming a Thirstier Resource? Excerpt from the World Energy Outlook 2012*. Paris: International Energy Agency.

Jackson, T. (2009) *Prosperity Without Growth? The Transition to a Sustainable Economy*. London: UK Sustainable Development Commission.

Jansz, S. and Wilbur, J. (2013) *Women and WASH: Water, Sanitation and Hygiene for Women's Rights and Gender Equality*, Briefing note. London: WaterAid and the Water Supply and Sanitation Collaborative Council.

Meadows, D., Meadows, D., Randers, J., and Behrens, W. (1972). *The Limits to Growth*, New York, NY: Universe Books.

Nordhaus, T., Shellenberger, M., and Blomqvist, L. (2012). *The Planetary Boundaries Hypothesis: a Review of the Evidence*. Oakland, CA: The Breakthrough Institute.

Nussbaum, M. (2003) 'Capabilities as fundamental entitlements: Sen and social justice', *Feminist Economics* 9(2–3): 35–59 <http://dx.doi.org/10.1080/1354570022000077926>.

Oosterlaken, I. (2013) *Taking a Capability Approach to Technology and Its Design: A Philosophical Exploration*. Delft: 3TU.Centre for Ethics and Technology.

Practical Action (2014) *Poor People's Energy Outlook Report*. Rugby: Practical Action Publishing.

Rawls, J. (1999) [1971] *A Theory of Justice, revised edn*. Boston, MA: Harvard University Press.

Raworth, J. (2012) *'A safe and just space for humanity: Can we live within the doughnut?'* Oxford: Oxfam.

Rockström, J., et al. (2009) 'Planetary boundaries: exploring the safe operating space for humanity', *Ecology and Society* 14(2): <http://www.ecologyandsociety.org/vol14/iss2/art32/>.

Rockström, J. (2012) 'Planetary boundaries: addressing some misconceptions' [online] Stockholm Resilience Centre <http://www.stockholmresilience.org/21/research/research-news/7-2-2012-addressing-some-key-misconceptions.html> [accessed 4 March 2016].

Rubenstein, M. (2012) 'Emissions from the cement industry', *State of the Planet*, New York: Earth Institute, Columbia University <http://blogs.ei.columbia.edu/2012/05/09/emissions-from-the-cement-industry/> [accessed 4 March 2016].

Sanderson, K. (2011) 'Chemistry: it's not easy being green', *Nature* 469: 18–20.

Schlesinger, W.H. (2009) 'Planetary boundaries: thresholds risk prolonged degradation', *Nature Reports Climate Change* <http://dx.doi.org/10.1038/climate.2009.93>.

Schulte, P., et al. (2013). 'Occupational safety and health, green chemistry, and sustainability: A review of areas of convergence', *Environmental Health* 12: 31 <http://dx.doi.org/10.1186/1476-069X-12-31>.

Sen, A. (1985) 'The standard of living', *The Tanner Lectures on Human Values delivered at Clare Hall, Cambridge University, March 11 and 12* <http://tannerlectures.utah.edu/_documents/a-to-z/s/sen86.pdf> [accessed 5 December 2015].

Steffen, W., et al. (2015) 'Planetary boundaries: guiding human development on a changing planet', *Science*, 347: 737–46 <http://dx.doi.org/ 10.1126/*science*.1259855>.

Stockholm Resilience Centre (2015) *Planetary Boundaries 2.0 New and Improved*, from Stockholm Resilience Centre: <http://stockholmresilience.org/21/research/research-news/1-15-2015-planetary-boundaries-2.0---new-and-improved.html> [accessed 5 December 2015].

The World Commission on Environment and Development (1987) *Our Common Future*, Oxford: Oxford University Press.

UN (2012) *Resilient People, Resilient Planet: A Future Worth Choosing*, New York: United Nations High-level Panel on Global Sustainability.

SE4All (2013) *Global Tracking Framework*, New York: United Nations Sustainable Energy for All Initiative.

Vinod, M.J. and Deshpande, M. (2013) *Contemporary Political Theory*, Delhi: PHI Learning Private.

White, S. (2009) *Bringing Wellbeing into Development Practice*, University of Bath Wellbeing in Developing Countries Working Paper 09/50. Bath: Bath University.

WWF (2011) *Draft Declaration on Planetary Boundaries*. Gland: World Wildlife Fund.

WWF (2012) *Living Planet Report*. Gland: WWF.

CHAPTER 3

Technology Justice and access to basic services

The light bulb – will it ever catch on? Access to energy services

We are living in a shadow under the light.

Resident of a village in Nepal in sight of a transmission line but with no electricity connection (author's recollection)

The injustice of energy poverty

Thomas Edison filed a patent for what was to become the first commercially viable incandescent electric light bulb on 4 November 1879. And yet today, nearly 140 years later, 1 billion people still live in the dark, with no access to electricity, while at least a further 1 billion people have only intermittent access to poor-quality electricity supplies (UN Foundation, 2015). In total, this constitutes about 30 per cent of global population or about 400 million households with very poor or non-existent access to a technology that is fundamental to the achievement of even a very basic standard of living.

As Figure 3.1 shows, the vast majority of the 1 billion without any access to electricity are located in the poorest countries of sub-Saharan Africa and South Asia. Around 87 per cent live in rural areas in low-density and relatively remote communities that are difficult and expensive to link to a national electricity grid. The population lacking access to electricity is dwarfed by that lacking clean cooking facilities, however. Around 2.9 billion people today still cook over open fires (SE4All, 2015), a population that is more evenly spread across rural and urban locations but which is still primarily located in the developing world.

Lack of access to modern energy services is a huge burden on those who experience it. Energy is needed in the home to provide light. A candle or simple wick kerosene lamp provides just around 11 lumens of light as compared to around 850 lumens from a basic 15-watt CFL fluorescent bulb (Practical Action, 2014), a clear illustration of why productive activities after dark are almost impossible without electricity. Energy needed to cook is usually the biggest household energy demand. Traditional open fires and simple solid fuel stoves are not only inefficient, meaning cooking takes much longer, but also create indoor air pollution responsible for over 4 million premature deaths a year among children and adults from respiratory and cardiovascular diseases, and cancer (WHO, 2014). The collection of fuelwood for cooking is also a

[DOI] http://dx.doi.org/10.3362/9781780449043.004

huge physical burden on the poorest – mostly on women and children – and on the environment. Alongside lighting and cooking, energy is, of course, also needed in the home to provide space heating in cold climates, to power refrigeration to preserve food, and to communicate with the outside world.

Energy access is not only a household-level issue. Electricity also enables people's livelihoods, whether it is to drive irrigation pumps or to power a lathe or welding machine in a workshop, or a freezer in a small cold store. Energy for heating and cooking plays an important role in small cafés and teashops, and also in food-processing activities that add value to local crops. Energy for mechanical processing is also frequently needed for small enterprises. The milling of grains, alongside the pressing of seeds for oil, or the removal of husks or shells, is one of the most common non-farm enterprise sectors. Cooling is used extensively in the food production chain for both storage and transportation. For example, milk chillers allow smallholder farmers to pool their milk production and store it until a tanker from a dairy can collect it. Transforming raw materials into end or intermediate products such as timber planks or wooden furniture can be done by hand, but is more efficient with powered tools, while repair of equipment often requires welding or powered workshop machinery. Information and communication technologies (ICTs) often spread quickly when electricity is available and can also have livelihood benefits. Television, radio, internet, and mobile phone services can bring more clients to a shop, bar, or restaurant.

In addition to providing services in the home and for livelihoods, access to energy, and to electricity in particular, is also important in the provision of community services, such as education, water supply, security, and health. To take the latter as an example, electricity means a rural health post can remain open after dark and that a refrigerator can maintain a cold chain for vaccines – two of the many energy services necessary to keep a basic health facility running, as summarized in Table 3.1.

The underlying causes of injustice

If the scale of the energy access problem today is an example of a technology injustice, it is an injustice that is compounded by the approach of the planners, regulators, governments, and financial institutions in charge of addressing it. The rural nature of the electricity access deficit means that technology choice plays a large role in determining who gets access to services. Extending national electricity grids is an expensive and relatively uneconomic business in low-density, remote, rural communities. But 'off-grid' alternatives exist. At the household level, for example, solar panels, batteries, LED lights, and other low-voltage appliances are options. At the community level, self-contained 'mini-grids' connecting households to a local network powered by a renewable energy source (micro-hydro, solar, wind, biogas, etc.), a diesel generator, or a hybrid of these technologies is also a possibility. Furthermore, these options are becoming more affordable, with the cost of solar panels in particular plummeting in recent

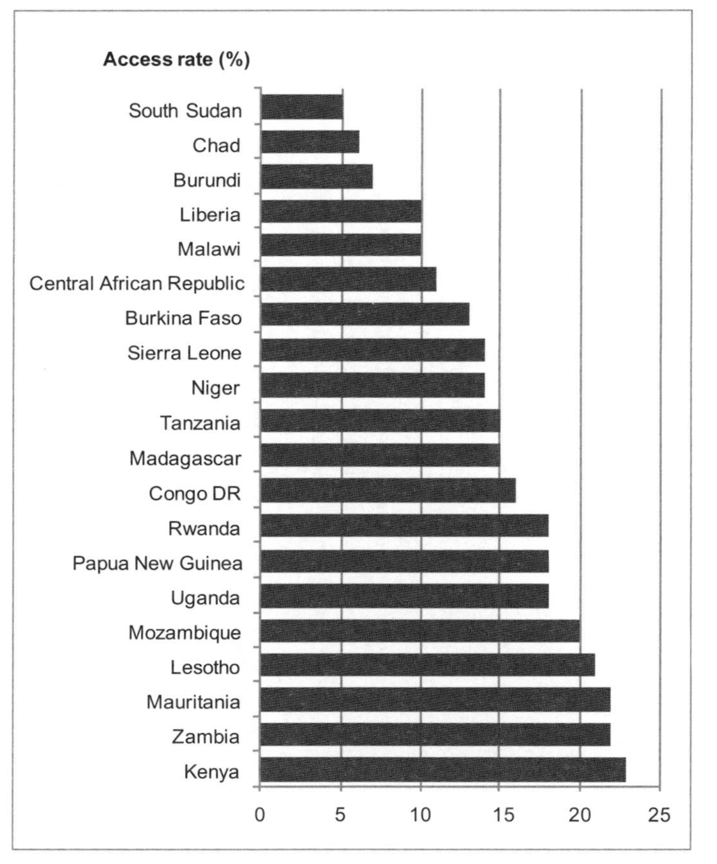

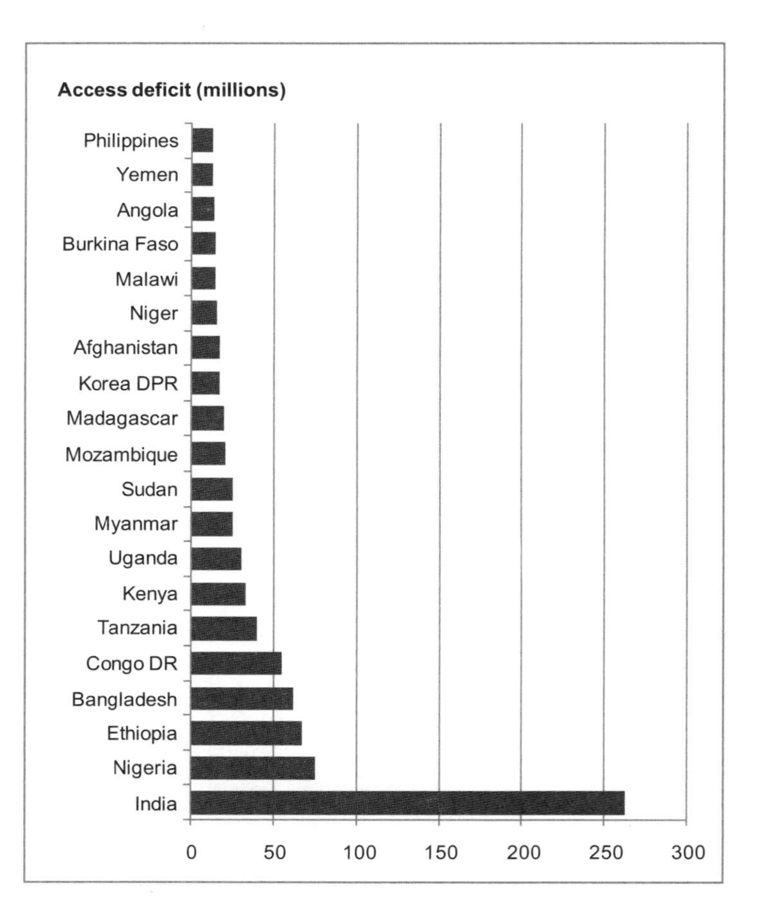

Figure 3.1 Top 20 electricity access deficit countries in proportion and absolute population terms
Source: SE4All, 2015: 48

Table 3.1 Energy services required for a small health post

Purpose/service	*Energy service/equipment*
General amenities/ infrastructure	
Basic amenities and equipment	Lighting – clinical/theatre, ward, offices/administrative, public/security Mobile phone charger, VHF radio, office appliances (computer, printer, internet router, etc.) Cooking, water heating and space heating Refrigerators, air circulation (electric fans) Sterilization equipment (dry heat sterilizer or an autoclave) Space heating
Potable water for consumption, cleaning, and sanitation	Water pump (when gravity-fed water not available) and purification
Health-care waste management	Waste autoclave and grinder
Service-specific medical devices	
Cold chain and Expanded Program on Immunization refrigeration	Vaccine refrigerator
Maternity and mother/child health	Suction apparatus, incubator, other equipment
HIV diagnostic capacity	ELISA test equipment (washer, reader, incubator)
Outpatient department	Portable X-ray, other equipment
Laboratory and diagnostic equipment	Centrifuge, haematology mixer, microscope, blood storage, blood typing equipment (37°C incubator and centrifuge), blood glucose meter, X-ray, ECG, ultrasound, CT scan, peak respiratory flow meter
Surgical equipment	Equipment and facilities for: tracheostomy; tubal ligation; vasectomy; dilatation and curettage; obstetric fistula repair; episiotomy; appendectomy; neonatal surgery; skin grafting; open treatment of fracture; amputation; cataract surgery
Additional infrastructure	
External lighting	Security lights at front gate, main doors, toilet block, etc.
Staff housing	Lighting, TV, AM/FM stereo, other appliances (mobile phone charger, electric fan, etc.), cooking and water heating
Emergency transportation	Vehicle or motorbike

Source: Practical Action, 2014: 36

years. The cost of off-grid electricity generation is rapidly approaching that of the historically cheaper grid technologies (see Figure 3.2).

The International Energy Agency has produced a costed scenario of how it might be possible to reach universal access to electricity and clean cooking services by 2030 (IEA, 2011). Two things stand out from this analysis:

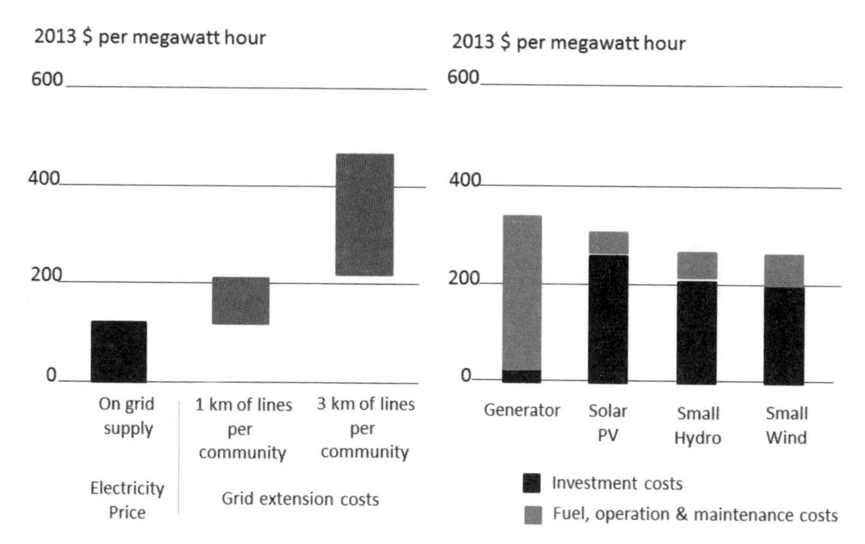

Figure 3.2 Indicative levelized costs of electricity for on-grid and off-grid technologies in sub-Saharan Africa, 2013

Source: SE4All, 2015: 62

- There is a massive funding deficit. Achieving universal access by 2030 will require increased investment in the sector from the current level of around $9 bn per annum to something in the order of $45 bn per annum.
- Because of the rural nature of the electricity problem, on average around 65 per cent of the additional investment necessary to provide universal electricity services will need to be in off-grid technologies, notably solar home systems or mini-grids (see Figure 3.3, which shows how this breakdown varies over the 20 years to 2030).

Shockingly, the proportion of the investment that needs to go into off-grid technology has turned out to be as big a problem as the scale of the overall investment required. Although it is difficult to analyse, as the sector doesn't track the split between on-grid and off-grid investment, a look at the lending patterns of the multilateral development banks (MDBs) over the period 2011–13 paints a picture that is broadly representative of the current investment climate (Sierra Club and Oil Change International, 2014). Figure 3.4 shows how far away the MDBs are from getting the split between on-grid and off-grid investment right; the African Development Bank, the MDB serving the continent with the greatest off-grid investment need, has made no investment at all in that technology over the three years under review.

There is a huge sectoral inertia to overcome if universal energy access is to be achieved. The formal energy sector is set up to build and operate big power stations and to deliver electricity through national grid systems. It struggles to confront the idea that what it has always delivered will not be the solution to ensuring universal energy access. Its engineers and technicians have been trained

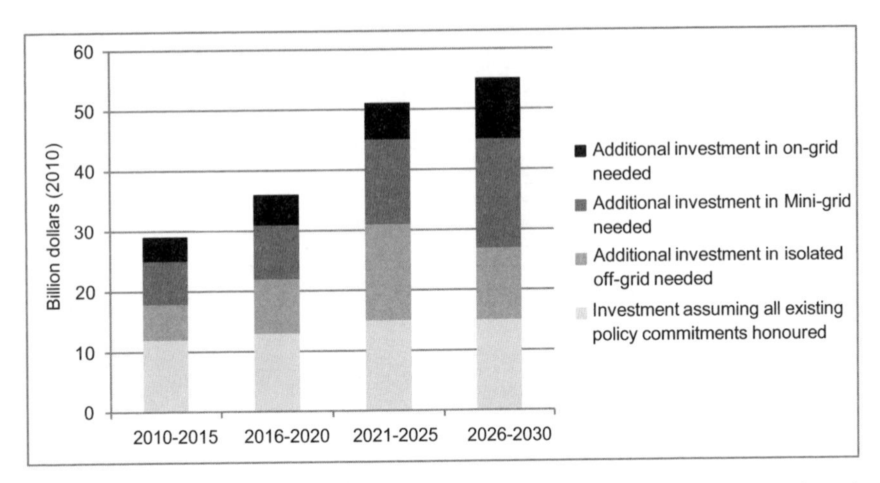

Figure 3.3 Average annual investment in access to electricity, by type, needed to meet universal access by 2030

Source: IEA, 2011: 22

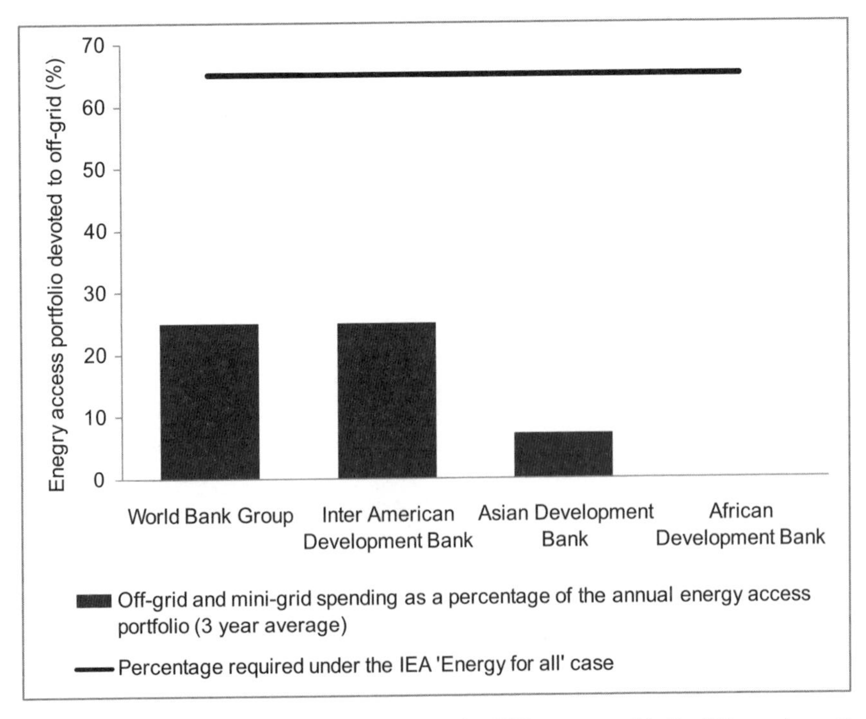

Figure 3.4 Lending for off-grid energy access by the MDBs compared to the IEA requirement to ensure universal access by 2030

Source: Based on data from Sierra Club and Oil Change International, 2014: 6

in grid technology and think in those terms, while its management structure is designed to administer the centralized system that a grid structure represents.

The engineers and technicians not only lack training in off-grid renewable energy technologies, but often see them as second best and far less attractive to work on. The administrative system struggles to work out how it might regulate, manage, and maintain large numbers of small, distributed power plants or how to engage with and bill the consumers who use its services. Just as importantly, the scale of typical grid extension and centralized power-generation projects lends itself more readily to the business norms of the international financial institutions than does the small-scale nature of mini-grids or household solar systems. The huge size of investments in grid, it must be noted, also provides more scope for leakage and corruption – perhaps another reason why some of those in control in the sector are so keen to promote grid over off-grid.

As a result of these institutional constraints, the formal energy sector largely washes its hands of responsibility for delivering off-grid energy services, leaving that to civil society, the private sector, or quasi-governmental bodies such as the Alternative Energy Promotion Centre in Nepal (responsible to the Ministry of Science, Technology and the Environment as opposed to the Ministry of Energy, which focuses solely on power provided through the grid). National energy planning becomes more difficult and the opportunity for urban and industrial consumers to cross-subsidize the tariffs of poor rural consumers is lost. This not only means that off-grid solutions attract less funding, but also that the subsidies generally present in the delivery of state-managed grid-based electricity are not enjoyed by those least able to pay – the rural poor, who are instead expected to finance a commercial rate of return to cover the costs of independent, private power producers investing in the infrastructure.

Finally, the way energy access is defined and measured in national statistics does little to push planning of services in the right direction. National statistics traditionally record the number of people with an electricity connection to the grid, but this is a poor measure of access to energy services for two reasons:

- It fails to recognize services provided through off-grid technologies, particularly solar home systems, stand-alone lighting such as solar lamps, and the adoption of clean and efficient cooking technologies. The failure to shine a spotlight on these areas in national statistics has the effect of deprioritizing them in national planning.
- It overestimates access in that the measure of having a 'connection' takes no account of the quality or cost of the electricity supply provided. A grid supply that is available only for a few hours a day, has wildly fluctuating voltage, or is too expensive for consumers to use for anything but basic lighting arguably does not provide energy access.

The United Nations Sustainable Energy for All (SE4All) initiative has tried to tackle this problem by proposing a tiered structure of indicators that better reflects what energy services a consumer is able to actually use (see Table 3.2).

A piloting of this measurement approach in the city of Kinshasa in the Democratic Republic of the Congo highlighted just how far traditional statistics

Table 3.2 Multitier matrix for access to household electricity supply

		Tier 0	*Tier 1*	*Tier 2*	*Tier 3*	*Tier 4*	*Tier 5*
Tiers	Tier criteria	–	Task lighting *and* phone charging	General lighting *and* television *and* a fan (if needed)	Tier 2 *and* any medium power appliances	Tier 3 *and* any high-power appliances	Tier 3 *and* any very high-power appliances
Appliances	*Indicative list of appliances*	–	*Very low-power appliances*	*Low-power appliances*	*Medium-power appliances*	*High-power appliances*	*Very high-power appliances*
	Lighting	–	Task lighting	Multi-point general lighting			
	Entertainment and communication	–	Phone charging, radio	Television, computer	Printer		
	Space cooling and heating	–		Fan	Air cooler		Air conditioner, space heater
	Refrigeration	–			Refrigerator, freezer		
	Mechanical loads	–			Food processor, washing machine, water pump		
	Product heating	–				Iron, hair dryer	Water heater
	Cooking	–			Rice cooker	Toaster, microwave	Electric cooking
Consumption	Daily consumption levels (watt-hours)	< 12	≥ 12	≥ 200	≥ 1,000	≥ 3,425	≥ 8,219

Source: SE4All, 2015: 213

can overestimate the level of energy access actually experienced. The formal sector statistics (using the binary metric of having or not having a connection) report that 90 per cent of the city's population has an electricity connection (and therefore assumed access). The multitier metric presents a very different picture. It is possible to calculate an energy supply index by looking at the proportion of the population with access to an electricity service equivalent to each of the six tiers on the scale in Table 3.2 and calculating an average. An index of 100 would have everyone on tier 5; an index of 0 would have everyone on tier 0. With an energy supply index of 30, Kinshasa's households thus have poor access to electricity, despite almost 90 per cent of them being nominally connected to the grid. In fact, nearly 80 per cent of households are on tier 2 or below and most of the remaining households are only on tier 3 (see Figure 3.5).

Achieving universal access

The global gap in energy access is an injustice because energy services are a prerequisite to achieving a basic standard of living and because the technology needed to address the gap is widely available now. It is an injustice compounded by the choices made by governments and international financial institutions about where to invest public funding. Continuing to favour grid-based (often fossil-fuelled) technologies over distributed (and often renewable) power production essentially prioritizes improving power supplies for industry and domestic customers who are already connected over providing basic energy services to those who have none. Decisions about how to encourage private-sector finance and where to place subsidies in the energy system also tend to favour consumers connected to the grid while those who are off-grid have to cover the full costs of their supply. To compound matters, national statistics are constructed in a way that exaggerates what has been delivered through the grid while ignoring much of what has been achieved by other means, consequently failing to provide the information that can drive change. Small wonder the IEA has concluded that, under current policies and rates of progress around the world, 'the absolute numbers of people without access to modern energy in 2030

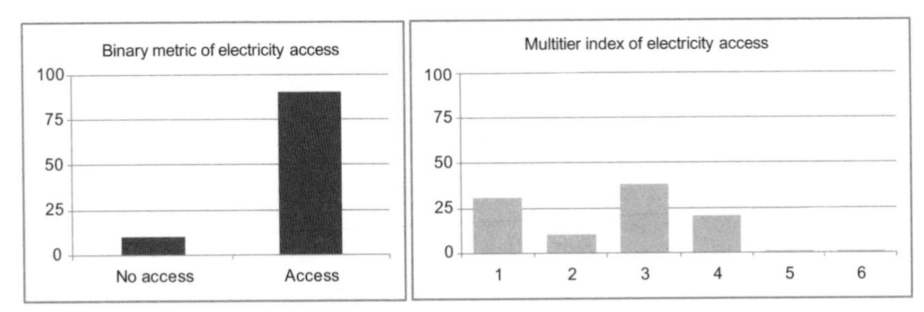

Figure 3.5 Kinshasa electricity access measured by connections and by energy services
Source: SE4All, 2015: 32

will be scarcely changed (though the proportion of the global population so deprived will have fallen). In sub-Saharan Africa, the numbers without modern energy access will have actually increased' (IEA, 2011: 43).

Things do not have to be this way. The technologies necessary to address the issue and the policy prescriptions to deploy them already exist. In order to move forward, three things need to happen.

Firstly, the scale and nature of the problem has to be recognized. If, as expected, the Global Tracking Framework of the SE4All initiative is implemented, its multitier matrix will help provide a much clearer picture of the existing state of access to energy services at national and global levels.

Secondly, because much of the solution is off-grid, many more stakeholders than just the formal energy sector need to be accounted for in national energy planning if universal access is to be achieved. These players will come from the private and civil society sectors as well as government and may include small independent operators of mini-grids, local manufacturers of improved cook-stoves, solar home-system installers, finance institutions (for both consumer and producer financing), importers and national distributors of technology such as solar panels and batteries, national standards bureaus (to enforce minimum-quality standards on off-grid equipment), customs and excise offices (to ensure appropriate tax relief on the import of renewable energy equipment), and, possibly, other government departments besides energy ministries. The skills and capabilities needed by all these different groups to effectively play their part in delivering off-grid energy solutions must be reviewed and, where necessary, strengthened. Existing policies and regulations will also need to be reviewed and strengthened to make sure they support and encourage, rather than obstruct, off-grid energy production.[2]

Thirdly, it will be necessary to address the finance gap at three levels:

- boosting the overall amount invested in the sector from $9 bn to around $45 bn per annum;
- shifting the balance of that funding away from grid-based technology to around 65 per cent for off-grid supplies;
- managing subsidy or other mechanisms to address the affordability gap for the poorest consumers, helping them move beyond lighting-only solutions to true energy access that includes all the energy services required in the home, for community services, and to power livelihoods.

It is perfectly feasible that, by 2030, every household in the world could have access to the light bulb, a technology first commercialized in 1879. But for the injustice of energy poverty to be addressed, significant changes have to be made in the technology choices of the energy sector and, consequently, the sector's institutions, policies, and financial instruments must also change dramatically.

The tap – a technology whose time has finally come? Access to water and sanitation services

Thousands have lived without love, not one without water.

W.H. Auden

The nature of an old injustice

If 140 years seems like a long time to wait for universal access to electricity, it is but a blink of an eye compared to the wait for universal access to safe water and safe sanitation. John Snow's report *On the Mode of Communication of Cholera* was published in London in 1849, but the link between water, sanitation, and disease had been understood long before the 19th century. The Romans had latrines for the safe disposal of human waste as well as piped water supplies, as the fragments of folded lead sheet pipes in the remains of their public baths attest. It is impossible to think of a more fundamental need or a service more central to the establishment of a social foundation than access to clean water and safe sanitation. And yet today, 2,000 years after the Romans and 165 years after Snow, 768 million people remain without a safe water supply and 2.5 billion people without safe sanitation.

Open defecation and untreated water supplies pose huge health risks that result in the deaths of around 800,000 children under five every year from diarrhoeal disease and cholera (Liu, 2012). Other serious diseases, such as typhoid fever, hepatitis, polio, legionellosis and leptosperosis, are also considered waterborne and so partly amenable to control through improvements to water quality and access to sanitation (UNICEF, 2008).

From a health perspective, increasing the quantity of water available is as important as increasing its quality, as a number of diseases are considered to be 'water-washed' because they are controlled by washing and improved personal hygiene. These include:

- soil-transmitted helminths (intestinal worms) such as *ascaris*, hookworm, and whipworm which, between them, infect around 25–33 per cent of the world's population. Worms suck blood and deprive their hosts of essential nutrients (particularly iron and Vitamin A) and over 130 million children suffer from high-intensity infections;
- trachoma, the world's leading cause of preventable blindness. About 6 million people are blind due to trachoma and more than 10 per cent of the world's population is at risk;
- ringworm, a fungal infectious disease of the skin and scalp;
- infections from fleas, lice, and ticks such as scabies and lice-born typhus (UNICEF, 2008).

Securing the desired health benefits from improved water supply and sanitation facilities is more complicated than it might first seem. The provision of clean water and safe sanitation to individual households does not in itself guarantee the elimination of waterborne and water-washed disease – a good example of technological determinism not playing out in practice. Studies have shown only modest reductions of 11–16 per cent in diarrhoeal disease from water and sanitation interventions alone, for example (WHO, 2014: ix). There are two reasons for this. Firstly, safe sanitation is a communal rather than an individual benefit. Having a safe means of disposing faecal matter in a particular household does not remove all sources of potential exposure to faeces. That only happens when all households in a neighbourhood also have and use latrines. Secondly, and most critically, human behaviour must also adapt to ensure the potential benefits of these technologies are actually realized. People have to adopt safe hygiene behaviours – use of the latrine, safe storage of water in the house to prevent its recontamination, and hand-washing with soap, particularly after defecation and before eating – in order to break the disease transmission route completely. Promoting hygiene behavioural change has its own challenges as health messages may be interpreted through local cultural understandings of health, illness, and the causes of disease (see, for example, Kaltenthaler, 1996).

Access to safe water is not only an issue of health. Like the collection of firewood for cooking, the collection of water for domestic use represents a huge physical burden and a drain on time, keeping children (especially girls) out of school and adults (mostly women) away from productive economic activity. In sub-Saharan Africa women collectively spend some 40 billion hours every year collecting water (UNIFEM, 2008).

As was the case for energy, access to safe water and, in some cases, sanitation services is also important for the provision of key social services and for productive purposes. Both are needed to maintain hygiene in schools and in health facilities; a domestic water supply is required in many rural areas to enable vegetable growing in kitchen gardens for a healthy diet; and water is also necessary for a wide range of income-earning enterprises, from hairdressing to food processing.

Lack of access to safe water and sanitation is primarily an issue for the poorest nations of the world. As Figure 3.6 shows, sub-Saharan Africa and Oceania fare far worse than other regions in access to clean water. The sanitation problem, however, is more evenly spread across the developing world and there are still 46 countries where less than half the population has access to improved sanitation.

But national coverage statistics do not tell the whole story, as they can hide great inequalities between populations within countries. Lack of access to these critical services is a burden borne predominantly by the poorest, usually rural, communities and those living in informal, low-income settlements in towns and cities. The inequality between urban and rural and poor and rich is most starkly demonstrated in sanitation, as Figure 3.7 shows for Mozambique.

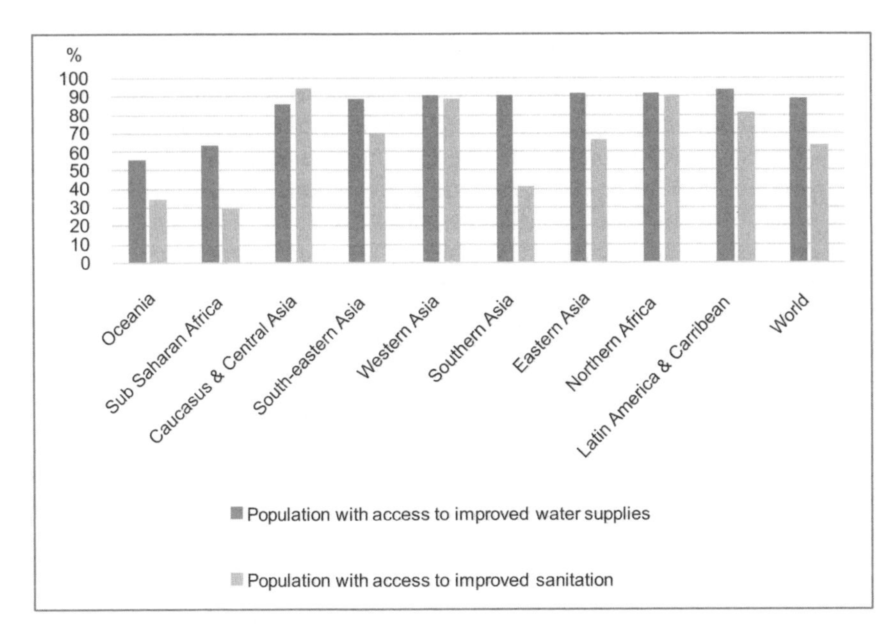

Figure 3.6 Proportion of population by region with access to improved drinking-water supplies or sanitation, 2012

Source: based on material from UNICEF and WHO, 2014

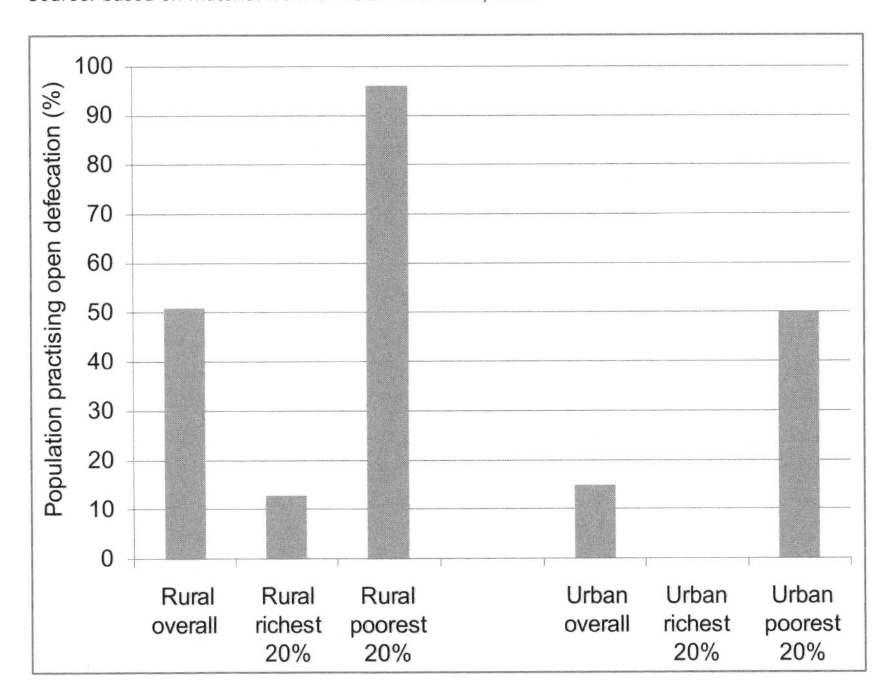

Figure 3.7 Inequalities between populations that practise open defecation in Mozambique

Source: based on material from UNICEF and WHO, 2014: 14

The special case of the urban poor – a lack of technology or of access rights?

In rural areas it is often a complete absence of water and sanitation infrastructure that is the major challenge in providing access. In urban spaces the challenge may not be establishing infrastructure from scratch, but about how existing infrastructure is shared between citizens. In urban centres the injustices around lack of access to safe water and sanitation services are particularly visible, with middle-class consumers connected to sewers and mains water living cheek by jowl with poor slum dwellers who have neither.

The reason many of the residents of slums in cities like Kenya's capital Nairobi lack access to safe drinking water is not because of an absence of technology – the mains water pipes are often already there, buried in the ground and providing supplies to their middle-class neighbours. Indeed, you have to walk over a large water main to enter parts of Kibera, Nairobi's largest slum. For residents of Kibera and other slums in the city, it is the informal nature of the settlements, unrecognized by government, that creates the barrier to accessing water and sanitation services. This lack of recognition excludes Kibera residents from entering into a supply agreement with the Nairobi water and sewerage utility. Despite their capacity to pay (discussed later), the availability of the necessary technology – often within a few metres of their house – and their evident need, a lack of formal land tenure rights trumps the slum residents' right to water. The key to change here therefore is not about the introduction of technology where there was none before, but about negotiating the right for certain excluded groups to access existing technology.

This is not an issue confined to the slums of Nairobi but one that is repeated across cities in Asia, sub-Saharan Africa, and Latin America. Globally, some 863 million people live in urban slums (UN Habitat, 2014). Many lack access to any form of safe sanitation and are forced to pay disproportionately high fees to purchase drinking water, often from unsafe sources provided via illegal and unregulated vendors.

Public investment choices as an underlying cause

Unlike energy, water and sanitation services were allocated targets to be achieved by 2015 under the Millennium Development Goals agreed at the United Nations in September 2000. The resulting attention to these services led to some significant progress being made, with almost 2 billion people gaining access to improved water supply and sanitation facilities since 1990 (UNICEF and WHO, 2014). Yet, although international aid for the water, sanitation, and hygiene sector increased by 30 per cent from 2010 to 2012, 70 per cent of 94 lower and middle-income countries in a survey reported that levels of financing were still insufficient to allow them to meet their national targets for drinking water and sanitation (UN-Water and WHO, 2014).

The injustice in provision of water and sanitation services does not stem from insufficient financial resources at a national level alone. The rural/urban and rich/poor discrepancies in access to these services, as shown in Figure 3.7, are not simply the result of inadequate national public finance but are also the consequence of decisions made about how the available public budget should be spent, decisions that could be made differently. The Global Analysis and Assessment of Sanitation and Drinking-Water (GLAAS) report, for example, shows that although the vast majority of people without access to basic sanitation live in rural areas, the bulk of financing continues to benefit urban residents, with expenditure on rural sanitation accounting for less than 10 per cent of total water and sanitation financing (UN-Water and WHO, 2014).

The lack of voice or political influence of the populations least able to access these basic services is, of course, a big part of the reason for this inequitable allocation of funding in national budgets. In addition, as in the energy sector, further challenges arise from the relatively low status and importance that engineers and technicians bestow on the small-scale distributed technologies which best meet rural needs (largely community-managed handpumps, small piped-water systems, and household latrines), compared to the big-ticket large infrastructure of urban water and sewerage grids and treatment plants. Hygiene promotion activities attract even less status and prioritization than small-scale decentralized infrastructure, despite the importance of behavioural change in securing health benefits from improved water and sanitation facilities. Only 19 of the 94 countries surveyed by the UN GLAAS report, for example, had an approved, funded, implemented, and reviewed national hygiene policy.

Action-oriented monitoring is another challenge for the sector. Although most countries have water and sanitation policies in place that contain specific measures for targeting the poor, far fewer actively monitor progress against those pro-poor targets. Furthermore, as the UN and World Health Organization note and Table 3.3 shows: 'only around 17 per cent

Table 3.3 Water and sanitation policy implementation of pro-poor governance, monitoring, and finance targets

	World Bank income category	*No. of countries*	*GOVERNANCE Universal access policy specifically includes measures for the poor (%)*	*MONITORING Monitoring system tracks progress in extending services for the poor (%)*	*FINANCE Finance measures to reduce disparity between the rich and poor are consistently applied (%)*
Sanitation	Low	32	81	38	12
	Middle	30	83	53	13
	Upper	26	73	35	27
Water	Low	32	81	41	22
	Middle	30	83	57	20
	Upper	26	73	42	27

Source: UN-Water and WHO, 2014: 16

of low and middle income countries have established and consistently apply financial measures that are targeted towards reducing inequalities in access to sanitation for the poor, and only 23 per cent for drinking-water' (UN-Water and WHO, 2014: 15).

Finally, in describing the underlying causes for this injustice arising from public investment policy, one must also consider a policy narrative which has taken root in recent years that water and sanitation services for the poor are a business opportunity or, to put it another way, that private-sector investment and full cost recovery are a means of accelerating the extension of such services to the poor. In rural areas this is particularly noticeable in sanitation and stems partly from a belief that sanitation (particularly on-site, such as a household pit latrine) is a private good, despite the obvious fact that the health benefits from stopping open defecation are clearly public (Mader, 2014). In urban areas it can be seen both in the provision of pay-per-use communal toilets as the only means of sanitation and in the relatively high proportion of slum and informal settlement dwellers who have to buy their water by the litre from vendors or kiosks as opposed to having a household piped connection (WHO and UNICEF, 2014).

Full cost recovery approaches to water and sanitation services for the poorest are often ineffective and unjust. Evidence from studies (for example, ODI, 2005; WaterAid; and London School of Hygiene and Tropical Medicine, 2008) shows that women and younger children are significantly less likely than men to benefit from pay-per-use sanitation facilities, partly due to their lack of access to and control over household cash income. Moreover, with the cash they do have, women prioritize spending on food, health, education, and other household essentials over spending on toilet access. Meanwhile, households purchasing water from vendors typically pay a higher price per litre than that paid by households connected to mains water supplies. This price differential or 'poverty tax' can be very large, with studies showing the poorest pay up to 7.5 times the rate per litre paid by middle-class consumers connected to official piped supplies in Kenya (UNDP, 2011); up to 15 times the standard piped rate in Ghana (Nyarko et al., 2008); and up to a staggering 37 times the official piped supply rate in Tanzania (Kjellen and McGranahan, 2006). These prices tend to be where water is purchased from unofficial vendors transporting it by truck or pushcart to areas without mains supply. Yet, even where prices are regulated through purchases from official water kiosks connected to the mains, the same studies show that the cost is at least around three to four times as much per litre as an official piped household connection. The result of full cost recovery approaches to expanding water and sanitation services for the poor can thus mean that the poor end up paying significantly more per unit of water or sanitation service than wealthier consumers who are able to obtain official household connections from utilities. This tax on poverty and the impact it has in denying those who cannot afford it access to technology that is critical to a minimum standard of life have to be considered unjust.

Achieving universal access

Like energy, the global gap in access to water and sanitation services is an injustice primarily because these services are a prerequisite to achieving a basic standard of living and because the technology needed to address the gap is widely available now. Again, like energy, it is an injustice compounded by the choices made about where to invest public funding by governments and international financial institutions that, by their nature, favour water supply over sanitation and hygiene promotion, urban over rural populations, and, within cities, official over informal settlements. These preferences have to be reversed and finance found to address the sanitation and hygiene promotion gap, which still leaves over a third of the world's population without a safe means of disposing of human excreta.

Achieving universal access to water and sanitation requires recognition that both services are essentially public rather than private goods, with widespread adoption required before the health benefits to society can be effectively achieved. Pricing and public subsidy policies need to reflect ability to pay and to reverse the 'poverty premium' paid by many of the poorest in society to access these services. South Africa, for example, addresses this for water by providing for free the first 6,000 litres a month for each household, but charging for all consumption over that amount, with unit costs rising as consumption increases beyond this level (Capetown Municipality, 2015).

Additionally, ways need to be found to allow water and sanitation utilities to operate in urban slums despite their informality, their sometimes temporary nature, and the issues of lack of land tenure and formal status that often complicate matters. The right to water and sanitation services needs to trump issues of land rights and there is evidence that this can be achieved. One example is the Mukuru settlement in Nairobi, where the NGO Practical Action worked with the Nairobi Water and Sewerage Company (NWSC) to formalize and improve water services to a low-income settlement. Like most slums there were no formal water supplies in the settlement at the outset. Instead, water vendors tapped the mains pipes illegally (and often in a very unsanitary manner) to take water for resale in small quantities to residents. The people of Mukuru paid a high price for a poor service, while NWSC had its pipes damaged and lost water for which it received no revenue. The solution lay in persuading the NWSC to reframe the problem from how to police informal settlements to reduce unaccounted water and financial losses, to how to serve a large group of potential customers who had already demonstrated a willingness to pay but were currently excluded from the service.

Achieving this was not as simple as it might sound because areas like Mukuru, as a result of their informality, can be difficult and sometimes dangerous places to operate in. NWSC staff were understandably sceptical about being able to enforce payment for a service and, in some cases, very nervous at the thought of having to enter the settlement to attempt this. Over time a workable approach evolved. The company agreed to extend official water supplies into parts of the settlement from existing take-off points from the main lines running along the periphery of the slum. The pipes and connections from the take-off points into the settlement were then managed and operated by water-user groups formed by the local communities, with NWSC billing each group for a bulk supply via water meters at the edge of the settlement. The water-user groups then distributed the water by a mixture of licensed connections to water vendors (who would typically have a reservoir tank and a tap stand from which to sell water by the jerry can) or household connections to those who could afford it. The vendors were made responsible for running repairs to the distribution lines and for collecting revenue and paying the bulk supply bill to the company. In return, water vendors could charge a regulated price of around 2 shillings per 20 litres (as opposed to the 5–10 shillings charged by unregulated sellers). The project aimed to legalize and regulate existing sellers rather than put them out of a job, and to encourage new sellers to extend services further where people could not afford household connections. The extent of the eventual transformation was significant and epitomized by the formation of a new department for informal settlements in 2008, with a managing director and a staff of engineers and sociologists that continues to work today (Nairobi Water and Sewerage Company, 2015).

Critical yet unavailable: access to essential medicines

> *Never have so many had such broad and advanced access to health care. But never have so many been denied access to health.*
>
> Gro Brundtland

Why 'essential' does not equal 'accessible' in the case of medicines

Over the past 100 years or so, technological developments in medicine, particularly in pharmacology and the production of vaccines, have transformed health care, reducing mortality rates and extending life spans. The expansion of immunization programmes over the past 40 years, which has seen 100 million children a year being immunized between 2005 and 2007, combined with access to important medicines such as antibiotics and antimalarial drugs, and improvements in basic services such as water and sanitation, has resulted in improvements in key indicators, among them mortality rates for children under five dropping from 17 million per year in 1970 to 9.2 million in 2007 (WHO et al., 2009), despite the global population rise over that period.

Slowly but surely, around the world acute diseases (communicable diseases such as malaria, tuberculosis, HIV/AIDS, measles) are being brought under control as a result of improved access to modern medical technologies in the form of vaccines and drugs. In their place chronic diseases (cancer, heart disease, etc.) are becoming the preponderant global health burden as lifestyles and diets change. But there are exceptions to this trend. In Africa, acute/infectious conditions are expected to continue to predominate until at least 2030 and possibly beyond. Nowhere is this difference between Africa and the rest of the world as stark as with the three infectious diseases tuberculosis, HIV/AIDS, and malaria, which form a much greater share of the overall health burden in Africa than elsewhere, as Figure 3.8 shows.

There is likely to be a number of reasons for this continued predominance of infectious diseases in the DALY burden in Africa, including the emergence of new drug-resistant variants of diseases such as tuberculosis and malaria on the continent. Much of the difference, however, must be a consequence of the relatively low level of access to medical facilities and medical technology for large parts of sub-Saharan Africa. Table 3.4 makes this comparison clear and shows Africa has around a tenth of the number of hospital beds per capita and less than half the number of doctors per capita than South-east Asia, the next lowest region in the WHO survey. Europe by comparison has over 60 times the number of hospital beds and 15 times the number of doctors per capita than Africa.

The problem of low numbers of medical staff and facilities is compounded by poor access to basic medicines necessary for the treatment of common

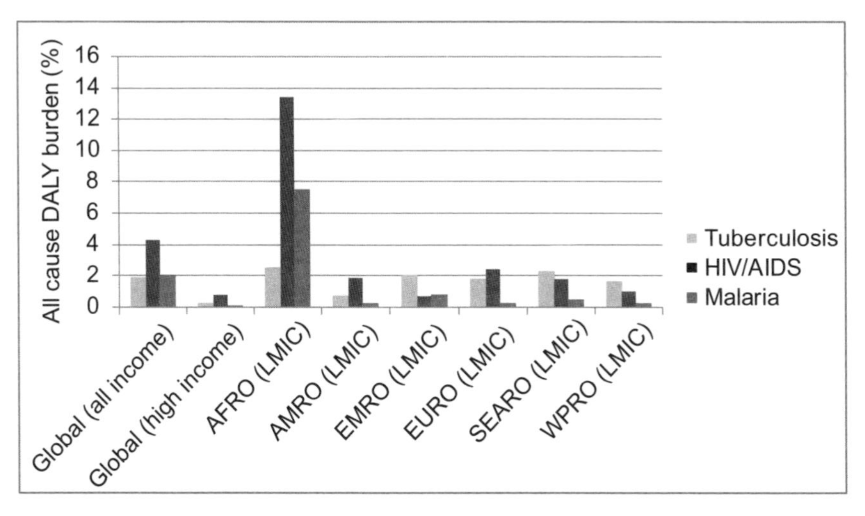

Figure 3.8 Disproportionate share of DALY burden from tuberculosis, HIV/AIDS, and malaria in Africa compared to other regions

Note: DALYs = disability-adjusted life years, the sum of years of potential life lost due to premature mortality and the years of productive life lost due to disability. One DALY can be thought of as one lost year of 'healthy' life.

The graph is broken down by WHO geographical regions: AFRO = Africa, AMRO = the Americas, EMRO = Eastern Mediterranean, EURO = Europe, SEARO = South-east Asia and WPRO= Western Pacific; LMIC = lower and middle-income countries.

Source: Kaplan and Mathers, 2011

Table 3.4 Availability of health services around the world

Region	*Hospital beds per 10,000 population*	*Doctors per 10,000 population*	*Nurses per 10,000 population*
Africa	<1	2.1	9.3
Americas	25	19.4	48.8
Eastern Mediterranean	13	7.4	11.1
Europe	64	32.0	74.3
South-east Asia	9	5.2	8.1
Western Pacific	31	11.0	17.0
World	*26*	*12.3*	*25.6*

Source: Peters et al., 2008

diseases. In 1977 the WHO defined a list of essential medicines on the basis that they 'satisfy the health care needs of the majority of the population ... [and] should therefore be available at all times in adequate amounts and in the appropriate dosage form' (WHO, 2003: 14). This list is updated every two years and around 95 per cent of developing country governments have published national essential medicine lists, often modelled on the WHO list and adapted to local circumstances (UN, 2008). But today the WHO estimates that around 30 per cent of the world's population has no access to essential medicines, with that figure rising to half the population in some areas (WHO, 2011).

Cost is clearly one barrier to access. Surveys in Kenya, for example, showed almost half of respondents reported problems finding cash to pay for treatment for their most recent illness. Means of coping included borrowing from friends, selling off possessions, buying smaller amounts of drugs than were prescribed, or simply not seeking care at all (Kanavos et al., 2011). In the developing world most medicine purchases are funded not by the state or health insurance but by out-of-pocket cash payments, meaning that serious illness comes with a high likelihood of households entering the debt and poverty cycle. This bears particularly hard on poorer households:

> Out-of-pocket spending is often proportional to the amount of care consumed and regressive, as usually it proportionately takes up large portions of lower income household budgets. Furthermore, there is no risk pooling or separation between risk of illness from financial risk. In a large number of developing countries, up to 90 per cent of the population purchase medicines on an out-of-pocket basis. In other words, medicines account for a significant proportion of personal or household income. This is in sharp contrast to most developed countries, where out-of-pocket expenditure for prescription medicines are a small proportion of total spending on medicines, due to health insurance coverage ... In the UK, for instance, the effective co-payment is 6 per cent, whereas in France and Spain it is 3.6 per cent and 7.8 per cent respectively. (Kanavos et al., 2011: 9)

Sheer availability of essential medicines is clearly a barrier to access, but the channel by which medicines are made available has an impact on their cost and so, again, their relative accessibility. As Figure 3.9 shows, even essential medicines tend to be more readily available through the private sector than the public: in 40 developing countries reviewed, essential medicines were available in the public sector in only one-third of cases but in the private sector they were available in two-thirds of cases (WHO cited in UN, 2008).

This differential availability has an impact on affordability and therefore accessibility because medicines provided through the public sector tend to either be subsidized or at least have a lower price mark-up than those provided by private health facilities and pharmacies. Low availability of medicines in the public sector therefore frequently forces patients to purchase medicines in the private sector, where prices are higher. A survey of 33 developing countries in 2008 showed a significant contrast in cost with, on average, the lowest priced generic medicines costing over six times what's known as the 'international reference price' (IRP) in the private sector compared to around just 2.5 times the IRP in the public sector (UN, 2008).

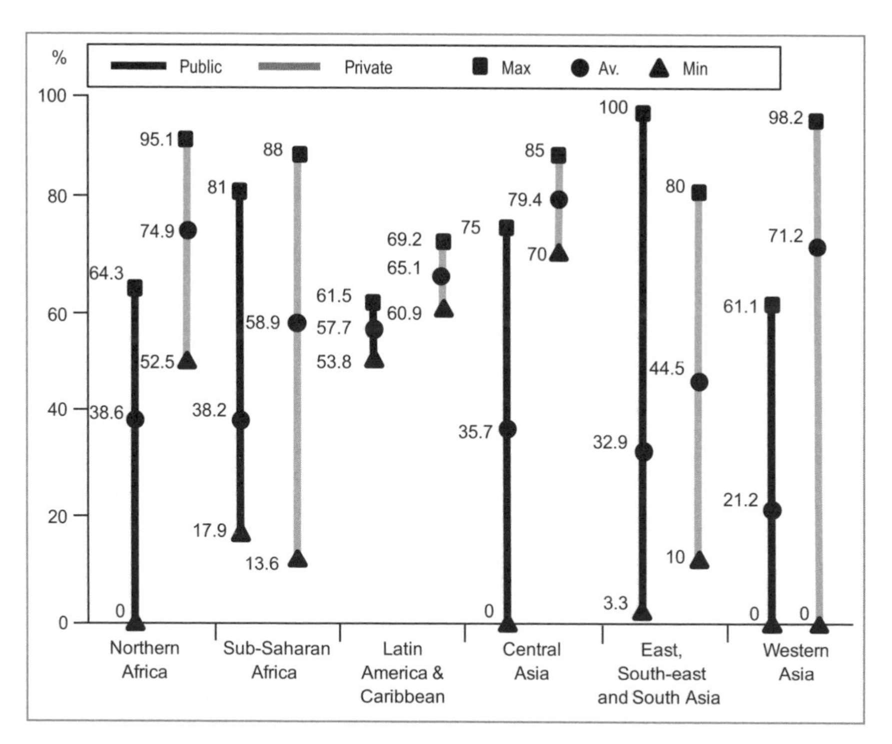

Figure 3.9 Availability of selected medicines in public and private health facilities, 2001–07
Source: UN, 2008: 37

International trade treaties and intellectual property rights

Taxes and duties on essential medicines, together with the efficiency of procurement and distribution processes and the presence or absence of policies to substitute generic for branded medicines where possible, all clearly have an impact on the cost to patients of medicines in the public sector. Similarly, price regulation affects prices in the private sector. But there is another factor that can also impact on cost: the rules on medicines and intellectual property rights under the Trade-Related Aspects of Intellectual Property Rights (TRIPS), as governed by the World Trade Organization.

Intellectual property rights (IPR) can take a variety of forms but, in the case of access to medicines, they generally involve patents. The purpose of patents is to stimulate innovation. Patents provide inventors with a monopoly and therefore an economic advantage. As such, governments hope that patent regimes create an incentive for research and development while, at the same time, delivering public benefits from technological advancement. Governments need to get the trade-off right between incentivizing innovation on the one hand, and ensuring that new products are widely available on the other. As Oxfam America notes:

> High levels of IP protection in developing countries exacerbate, rather than help solve, the problem of access to affordable medicines. Extensive patent protection for new medicines delays the onset of generic competition. And because generic competition is the only proven method of reducing medicine prices in a sustainable way, such high levels of IP protection are extremely damaging to public health outcomes. (Oxfam America, 2015)

There is some debate about how much IPR impinges on the accessibility of essential medicines in the developing world as many of the medicines on the list are long-established and not under patent, an almost self-fulfilling prophecy given that one of the criteria for selection on the WHO list is affordability (Watal, 2000). Indeed, a 2004 study suggested that only 17 of the 319 products on the essential medicines list at that time were patentable and even fewer patents had been enacted (Attaran, 2004). But with new drugs being added to the 'essential' list in 2015, including entirely new treatments for hepatitis C for the first time, as well as new, improved drugs for treatment of various cancers and tuberculosis, the issue of patents impacting on affordability of essential medicines does not look like it will go away (WHO, 2015).

In the early 2000s, HIV/AIDS focused attention on patents and prices. At the time an estimated 95 per cent of those suffering from this disease lived in developing countries and, as it was a relatively new disease, many of the medicines were also new and therefore still under patent protection. The struggle in South Africa to allow HIV-positive people access to cheap generic antiretroviral drugs and effective care is one of the best documented examples of the impact of IPR on access to essential medicines, although the story involves politics and the beliefs of senior political figures as well. The struggle centred on two barriers to poor people accessing the science-based technologies (in this case modern drugs) that would help them. One barrier was the refusal of President Thabo Mbeki to acknowledge the link between HIV and AIDS, which prevented the South African health service from adequately addressing an epidemic that, by 2007, was estimated to affect 5.4 million people (Heywood, 2009). The other barrier was a legal battle in 1998 between the South African government and a number of multinational drug companies over whether a section of the country's 1997 Medicines Act (designed to allow the manufacture of low-cost generic versions of antiretroviral drugs) contravened TRIPS rules. A well-publicized civil-society campaign, known as the Treatment Action Campaign and supported by international agencies such as Oxfam and WHO, created widespread support for the government's legal case and significant adverse publicity for the 39 drug companies involved, which were shamed into withdrawing from the court case in 2001 (Barnard, 2002). Even after the legal victory, however, it took a further two years before cheaper antiretrovirals became available, principally because of Mbeki's stance and the reluctance of the government to declare the AIDS epidemic as a national public health emergency, which would have then cleared the way for the import or local manufacture of generics under TRIPS anyway. The eventual change of government in 2003 subsequently removed this political barrier, leading to a (projected) 10-fold increase in the national health budget for HIV/AIDS over the period 2003 to 2010 (Heywood, 2009: 26). By 2008 nearly half of South Africa's HIV/AIDS-infected population was receiving antiretroviral drugs – still not ideal, but a big improvement.

Improving access to essential medicines

Technological advances in medicine in recent decades have made a huge difference to the quality of life of billions of people. The WHO list of essential medicines sets out the set of drugs that should be universally available to ensure these benefits are extended to everyone, at an affordable price. This is critical to the establishment of a social foundation, but access remains a problem for nearly 2 billion people in the world, despite the medicines on the list being overwhelmingly well established and largely unaffected by patent issues. The injustice is that, once again, the 2 billion affected are largely rural populations and the urban poor in the developing world.

Clearly, there are interlinking factors that affect access to essential medicines, including the availability of basic health infrastructure and trained staff to provide medical diagnoses and prescriptions, particularly in rural areas, and the many complex drivers of cost. Beyond the allocation of additional funding for health-care personnel and equipment in rural areas where local medical facilities are sparse, a recent technological approach has been to use video conferencing and sensor technology to allow remote consultation and diagnosis, which is discussed in more depth in Chapter 4.

But making the drugs on the WHO essential list widely available and affordable is critical to creating a fair distribution of opportunity to access the benefits of this technology. One approach to achieving this is differential pricing, where multinational pharmaceutical companies agree to, typically, three levels of pricing that are matched to lower, middle, and upper-income countries' ability to pay. This is not so much a global cross-subsidy as a recognition by companies (perhaps due to pressure to act in a more socially responsible manner) that they will make profit from lower-income countries only via high volumes of sales at low price points. To date the approach has been used most often for vaccines, where demand volumes are very high and, at least for vaccines aimed at tropical diseases, there is less potential for cheaper versions sold in low-income countries to leak back into developed country markets and so undermine prices there. In this model, full market prices are charged to developed countries, low prices to countries belonging to the Global Alliance for Vaccines and Immunization, and intermediate prices to middle-income countries. Alongside vaccines, differential pricing has been applied to antiretroviral drugs and to contraceptives, but is not widely seen in other medicines or medical services (Yadav, 2010).

More broadly, a review of progress against the Millennium Development Goals provides a good summary of the changes that are needed to ensure universal access to the WHO essential medicines list (UN, 2008: 43–44). At the national level these are to:

- eliminate taxes and duties on essential medicines;
- update national policy on medicines;
- update the national list of essential medicines;
- adopt generic substitution policies for essential medicines;
- seek ways to reduce trade and distribution mark-ups on prices of essential medicines;
- ensure adequate availability of essential medicines in public health-care facilities;
- regularly monitor medicine prices and availability.

At the global level these are to:

- encourage pharmaceutical companies to apply differential pricing practices to reduce prices of essential medicines in developing countries where generic equivalents are not available;
- enhance the promotion of the production of generic medicines and remove barriers to uptake;
- increase funding for research and development in areas of medicines relevant to developing countries, including children's dosage forms and most neglected diseases.

Technology Justice and basic services

As Raworth's analysis suggests, access to basic services such as water and sanitation, energy supplies, and essential medicines are critical to the establishment of a minimum universal social foundation – that lower boundary of a safe, inclusive, and sustainable space for human development. As the discussion in this chapter has revealed, provision of those services is dependent upon access to a suite of technologies that is already available and which is in widespread use in many parts of the world, but from which a substantial minority of the world's population has been excluded. This is technology injustice on a grand scale.

On the face of it, much of this injustice results from the differing abilities of poorer and wealthier sections of society to afford the costs of accessing these technologies. But the issue of affordability itself is exacerbated by national policies that result in perverse effects, ranging from the imposition of import duties on renewable energy equipment, making solar home systems for rural populations unnecessarily expensive, to cost recovery policies for urban water supplies that leave poor slum dwellers paying a higher unit cost for a poorer service than their middle-class neighbours receive.

Decisions around the allocation of public finance also have an important impact on the accessibility of critical technologies. The favouring of large-scale grid-based technology over distributed off-grid supplies in national decisions around public investment, whether for energy or water and sanitation services, inevitably results in further improvements to services for people who already have them over provision of first-time access for those who do not.

Not all factors underpinning these technology injustices can be blamed on national-level decisions by government, however. The tendency for multilateral development banks and bilateral donors to distribute funding in large blocks to minimize transaction costs also favours large-scale, centralized technology over the distributed forms more likely to deliver services to marginalized rural communities. The IPR and pricing policies of multinational pharmaceutical companies determine the cost of some medicines on WHO's essential list. A preference among donors to fund water supply over sanitation leaves access to the latter trailing behind the former.

Achieving Technology Justice in access to basic services will require action on all these fronts, together with better and more transparent systems for monitoring the true scale of the impact on poor people of failing to deliver these critical areas of the social foundation.

Notes

1. See the World Bank's RISE index for examples of policies to support off-grid and renewable energy generation (World Bank, 2016).
2. Based on a lifeline minimum consumption requirement of 25 litres per person per day.

References

Attaran A. (2004) 'How do patents and economic policies affect access to essential medicines in developing countries?', *Health Affairs* 23(3): 155–66 <http://dx.doi.org/10.1377/hlthaff.23.3.155>.

Barnard, D. (2002) 'In the High Court of South Africa, case no. 4138/98: the global politics of access to low-cost AIDS drugs in poor countries', *Kennedy Institute of Ethics Journal*, 12(2): 159–74 <http://dx.doi.org/10.1353/ken.2002.0008>.

Capetown Municipality (2015) *Cape Town water and sanitation tariffs 2015–2016* [online] <http://www.capetown.gov.za/en/Water/Documents/Water%20and%20Sanitation%20Tariffs%202015%202016.pdf> [accessed 19 January 2015].

Heywood M. (2009) 'South Africa's Treatment Action Campaign: combining law and social mobilization to realize the right to health', *Journal of Human Rights Practice*, 1(1), 14–36 <http://dx.doi.org/10.1093/jhuman/hun006>.

IEA (2011) *World Energy Outlook 2011, Energy for All: Financing Access for the Poor. Special Early Excerpt of the WEO 2011*. Paris: International Energy Agency.

Kaltenthaler, E.C. and Drašar, B. (1996) 'The study of hygiene behaviour in Botswana: a combination of qualitative and quantitative methods', *Tropical Medicine and International Health,* 1(5): 690–98 <http://dx.doi.org/10.1111/j.1365-3156.1996.tb00097.x>.

Kanavos, P., Das, P., Durairraj, (V.,) Laing, R., and Abegunde, D.O. (2011) 'The world medicines situation 2011 – Options for financing and optimizing medicines in resource-poor countries', in *World Medicines Situation Report 2011*, 3rd edn, Geneva: WHO <http://apps.who.int/medicinedocs/en/m/abstract/Js20033en/>.

Kaplan, C. and Mathers, W. (2011) 'The world medicines situation 2011 – Global health trends: global burden of disease and pharmaceutical needs', in *World Medicines Situation Report 2011*, 3rd edn, Geneva: WHO <http://apps.who.int/medicinedocs/en/m/abstract/Js20036en/>.

Kjellen, M. and McGranahan, G. (2006) *Informal Water Vendors and the Urban Poor*, London: Institute for Environment and Development.

Liu, L., et al. (2012) 'Global, regional, and national causes of child mortality: an updated systematic analysis for 2010 with time trends since 2000', *Lancet*, 379: 2151–61 <http://dx.doi.org/10.1016/S0140-6736(12)60560-1>.

Mader, P. (2014) 'Five myths of the World Bank's approach to water and sanitation', *Governance Across Borders*, <http://governancexborders.com/2014/02/05/five-myths-of-the-world-banks-approach-to-water-and-sanitation/> [accessed 18 January 2016].

Nairobi Water and Sewerage Company (2015) 'Brief on the informal settlements', <http://www.nairobiwater.co.ke/projects.html> [accessed 23 May 2015].

Nyarko, K., Odai, S.N., Owusu, P.A., and Quartey, E.K. (2008) 'Water supply coping strategies in Accra', *Proceedings of the 33rd WEDC International Conference, Accra, Ghana, 2008: Access to Sanitation and Safe Water: Global Partnerships and Local Actions*, Loughborough: WEDC, University of Loughborough.

ODI (2005) 'Livelihoods and gender in sanitation, hygiene and water services among the urban poor', London: Overseas Development Institute, Practical Action and the University of Southampton.

Oxfam America (2015) 'Intellectual property and access to medicine', <http://policy-practice.oxfamamerica.org/work/trade/intellectual-property-and-access-to-medicine/> [accessed 24 May 2015].

Peters, D., Garg, A., Bloom, G., Walker, D.G., Brieger, W.R., and Rahman, M.H. (2008) 'Poverty and access to health care in developing countries', *Annals of the New York Academy of Sciences*, 1136: 161–71 <http://dx.doi.org/10.1196/annals.1425.011>.

Practical Action (2014) *Poor People's Energy Outlook*, Rugby: Practical Action Publishing.

Sierra Club and Oil Change International (2014) 'Failing to solve energy poverty: how much international public investment is going to distributed clean energy access?' San Francisco and Washington: Sierra Club and Oil Change International.

UN (2008) *Delivering on the Global Partnership for Achieving the Millennium Development Goals: Millennium Development Goal 8*. New York: United Nations.

UNDP (2011) 'Small-scale water providers in Kenya: pioneers or predators?' New York: United Nations Development Programme.

UN Foundation (2015) 'What we do: achieving universal energy access', <http://www.unfoundation.org/what-we-do/issues/energy-and-climate/clean-energy-development.html> [accessed 22 April 2015].

UN Habitat (2014) 'World Habitat Day background paper', United Nations Human Settlements Programme, Nairobi, Kenya, <http://unhabitat.org/wp-content/uploads/2014/07/WHD-2014-Background-Paper.pdf> [accessed 16 March 2016].

UNICEF (2008) *UNICEF Handbook on Water Quality*. New York: United Nations Children's Fund.

UNICEF and WHO (2014) *Progress on Drinking Water and Sanitation Update 2014*. Geneva: World Health Organization.

UNIFEM (2008) *Progress of the World's Women 2008/2009: Who Answers to Women? Gender and Accountability*, New York: UNIFEM (now UN Women).

SE4All. (2015). *United Nations Sustainable Energy for All Global Tracking Framework*. New York: UN.

UN-Water and WHO (2014) *Global Analysis and Assessment of Sanitation and Drinking-Water: Investing in Water and Sanitation, Increasing Access, Reducing Inequalities*. Geneva: WHO.

Watal, J. (2000) *Access to Essential Medicines in Developing Countries: Does the WTO TRIPS Agreement Hinder It?* Cambridge, MA: Centre for International Development, Harvard University.

WaterAid and London School of Hygiene and Tropical Medicine (2008) 'Communal toilets in urban poverty pockets: use and user satisfaction associated with seven communal toilet facilities in Bhopal, India', London: WaterAid.

WHO (2003) *The Selection and Use of Essential Medicines*, WHO Technical Report Series no. 914, Geneva: WHO.

WHO (2011) *The World Medicines Situation 2011: Access to Essential Medicines as Part of the Right to Health*. Geneva: WHO.

WHO (2014) *Household Fuel Combustion: WHO Guidelines for Indoor Air Quality*. Geneva: WHO.

WHO (2015) 'WHO moves to improve access to lifesaving medicines for hepatitis C, drug-resistant TB and cancers', 8 May, <http://www.who.int/mediacentre/news/releases/2015/new-essential-medicines-list/en/> [accessed 7 March 2016].

WHO, UNICEF and World Bank (2009) *State of the World's Vaccines and Immunization*, 3rd edn. Geneva: WHO.

World Bank (2016) 'Readiness for Investment in Sustainable Energy', <http://rise.worldbank.org/> [accessed 15 January 2016].

Yadav, P. (2010) *Differential Pricing for Pharmaceuticals: Review of Current Knowledge, New Findings and Ideas for Action*, London: Department for International Development.

CHAPTER 4
Technology Justice and access to knowledge

The previous chapter considered how the provision of basic services supports a minimum standard of living, together with the need to correct injustices around access to technologies, such as water pumps, solar panels, and medicines, if Raworth's universal social foundation is to be achieved. But the 11 elements of the social foundation, or any similar construct of a universal minimum standard of living, will not be met without people being able to access and make use of 'soft' technological knowledge, as well as accessing 'hard' technologies, such as pumps and generating equipment. This chapter looks at one such area of critical technical knowledge – agriculture – and the injustices that need to be challenged there, before considering the role digital technology has to play in facilitating broader access to technical and other knowledge essential for establishing the social foundation.

Feeding the world: smallholder farmers' need for better access to technical knowledge

> *When you give food to the poor, they call you a saint. When you ask why the poor have no food, they call you a communist.*
>
> Archbishop Helder Camara, three-time Nobel Peace Prize nominee

Farming in extremis: the power of new knowledge

In the Gaibandha district of Bangladesh a modest agricultural revolution has taken place. Bangladesh is set at the confluence of two major river systems, the Ganges and the Brahmaputra, and the country is, in effect, one of the largest river deltas in the world. Every year the monsoon floodwaters from the Himalayas spread out across the country, with one-third of the land surface under water in a normal wet season and up to two-thirds in exceptionally wet years. Viewed from the air in the dry season, Bangladesh is a tapestry of braided river channels. Viewed over a number of years, the tapestry changes as each successive monsoon flood cuts new channels and fills in old ones, causing the braids to snake from side to side over time. Not only do the river channels change course but islands in the rivers (known as *chars*) gradually migrate downstream over the years, as the annual floodwaters erode soil from the upstream ends of the islands and deposit it again at the downstream tips.

http://dx.doi.org/10.3362/9781780449043.005

The monsoon floods bring nutrients and fertile silt to the plains of Bangladesh, enabling intensive agriculture with high productivity as a result. But although the monsoon is a blessing for most farmers, it is a curse for some. The heavy erosion along the river channels means that every year many farmers lose land as rivers switch course or islands erode. Sometimes this loss can be catastrophic and farmers lose all their productive land. Embankments have been built along the edges of many rivers to protect farms, but the scale of the flooding means it is neither possible nor desirable to contain the rivers along their whole length.

Families who have lost everything to the floods often end up in small houses made from bamboo mats and grass, perched on the side of flood embankments – the only safe public space left to build. The men of the family are often absent, having to leave to search for paid labouring work elsewhere since their families can no longer grow their own food. Those living on the embankments are some of the poorest and most deprived people in what is already a poor country. But some of the embankment dwellers in Gaibandha are now growing their own food again and producing a surplus for sale, even though they have no land. They are doing this by using a combination of two techniques: floating gardens and sandbar pitting.

Floating gardens are rafts woven from water hyacinths, a strong floating vine-like weed found in abundance in rivers. These rafts, often up to 2 metres wide and 10 metres long, are topped with soil into which vegetables are planted to provide a food supply during the monsoon. When the floods come, the rafts are tethered and simply float, rising with the water. People continue to tend these floating gardens during the floods, often standing waist deep in water. Once the floods recede and the bare sandy beds of the rivers are exposed again, the rafts are quickly turned into compost for the second part of this cultivation cycle, sandbar pitting. The trick here is turning sterile sandbars into fertile gardens. Small pits are dug in the exposed sandbars. The pits are then filled with the composted material from the rafts and used to grow further crops, mostly pumpkin, which has the added value of being a commercial crop and a vegetable that can be dried and stored for household use later in the year. During the six years from 2009 to 2015 nearly 13,000 landless farmers in Gaibandha took up the sandbar cultivation approach, producing over 54,000 tonnes of pumpkins during the period.[1] The technique has been transformational for those without land, providing food and the possibility of an income once more without having to migrate elsewhere.

How did this transformation come about? Through the simple act of making knowledge of the floating garden and sandbar pitting techniques, which were already in practice in the southern part of Bangladesh, available to the landless farmers in the northern district of Gaibandha.

Improving food security and livelihoods

It is difficult to underestimate the importance of agriculture to the economies of the developing world. In 2008, agriculture, on average, accounted for 29 per cent of gross domestic product (GDP) in low-income countries, compared to just 10 per cent for middle-income and 1 per cent for high-income countries (Bientema and Fuglie, 2012). Around a billion people are undernourished today (FAO, 2009); of these, 60 per cent are women and 25 per cent are children (FAO, 2006). The vast majority of these people are connected to food provision in some way: three-quarters are in rural communities (Bhatkal et al., 2015) where agriculture provides a livelihood for nearly 90 per cent of people (World Bank, 2007). Many of those who go hungry are therefore either food providers themselves or live in rural food-producing communities.

Sub-Saharan Africa epitomizes these links between food producers and malnourishment. Of the population of around 800 million people, around 223 million are currently undernourished, and this figure could rise rapidly over the next 30 years due to climate change and a projected growth in the population to 1.5 billion by 2050. Smallholder farmers are the main producers of food on the continent, accounting for 80 per cent of all the farms. They employ around 175 million people directly, 70 per cent of whom are women. Most work in degraded or poor soils and practise a subsistence form of agriculture with low inputs (including little or no irrigation) and low yields. Incomes are low, typically under $2 per day, with 60 per cent of income spent on food (AGRA, 2014).

It is not only in Africa that food provision is dominated by small-scale providers. An estimated 70 per cent of the global population, or nearly 4.7 billion people, are fed with food provided locally, mostly by small-scale farming, fishing, or herding (ETC Group, 2009: 1). Of the world's farms, 85 per cent are holdings of less than 2 hectares, worked by families and indigenous peoples. Frequently quoted figures place the number of small-scale farmers at 1.5 billion people (ETC Group, 2009). The importance of small-scale agriculture in securing the world's food is therefore clear and has been readily acknowledged (see, for example, the OECD's work on promoting pro-poor growth in the agriculture sector (OECD, 2006: 31)). But rural households also form about 75 per cent of the world's hungry people (Pinstrup-Andersen and Cheng, 2007). Often, they attempt to produce enough food for themselves and for local markets in conditions that are harsh and with little external support. It is predominantly women who farm smallholdings to feed their families. In the absence of effective storage facilities, and under pressure to raise income, excess produce is sold shortly after harvesting, achieving whatever price is available from local buyers at that time. Low cash income means that farming is, by default, often undertaken without chemical fertilizers and using seeds that are stored for the following year or exchanged with neighbours. What little money that is made is often used to buy produce at market, at higher prices, to supplement household food supply (De Schutter, 2009).

Access to knowledge on improved farming techniques could be as important and transformational for these smallholder farmers as it was for those in Gaibandha in Bangladesh. For these resource-poor farmers, improving soil fertility and moisture content, managing threats of pests and diseases, and sustainably increasing productivity with minimal external inputs, could make a major contribution to improvements in food security and rural livelihoods.

The state of agricultural extension services

Access to technical knowledge may be critical to improving food security and rural livelihoods, but there are major weaknesses in the agricultural extension services that are supposed to provide this advice. Government-funded advisory services were severely cut back during the 1990s in many developing countries in response to structural adjustment programmes and the resulting public spending cuts. Many have never recovered. Although government extension services still dwarf those of the private sector (see Figure 4.1), in reality their outreach is severely limited. A review of seven countries in Africa found the 'proportion of farmers seeing extension officers varies from just 1.3 per cent in Nigeria to 23 per cent in Zambia (but half of those see extension officers only "rarely")' (ActionAid, 2013: 10). Likewise, in India only 91,288 of the 143,863 positions in the Department of Agriculture were filled, which meant that, given the huge number of farm households in the country, extension services reached an average of only 6.8 per cent of farmers (GFRAS, 2012).

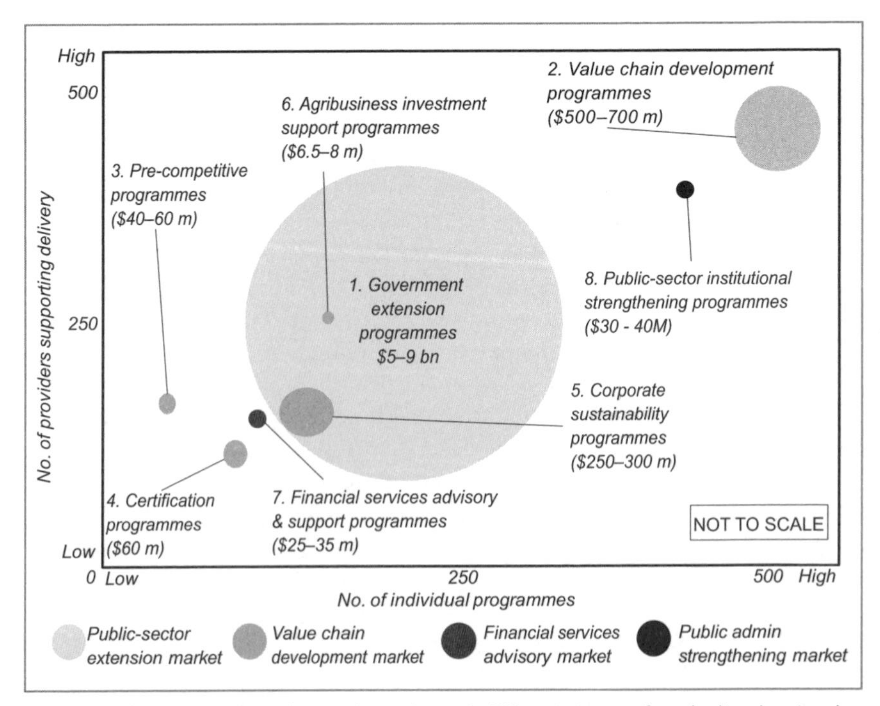

Figure 4.1 Relative global size and number of different types of agricultural extension programme

Source: Shakhovskoy, 2014

Figure 4.1 shows there is a number of alternatives to government-funded extension services, although they have very limited outreach by comparison. Value chain development programmes are typically funded by donors of international development assistance and work to improve productivity and to create market linkages for cash and some staple crops. These advisory programmes can include training in improved farming practices for a specific crop along with business and finance skills. Pre-competitive platforms are sometimes funded by donors such as the Gates Foundation to bring a group of industry players together to invest in the development of a specific market that is at a very early stage. Private sector-funded advice, meanwhile, tends to be tied to specific crops or inputs. Certification and corporate sustainability programmes are often funded by large multinational buyers and food processors to improve quality and increase the sustainability of their own specific supply chains, while agribusiness investment support programmes are managed by development finance institutions and other financiers to support agribusiness investments, sometimes with supplementary technical assistance. Generally speaking, private-sector initiatives focus more on areas of higher productivity with better soils and the potential for irrigation rather than on the marginal lands that most smallholder producers occupy.

There is another, more deep-seated problem with services delivering technical advice to farmers: they are almost entirely focused on men. As the Global Forum for Rural Advisory Services notes:

> Almost all extension services lack something crucial – female participation is very low. Women, on average, comprise 43 per cent of the agricultural labour force in developing countries and account for an estimated two-thirds of the world's 600 million poor livestock keepers. Yet only 15 per cent of the world's extension agents are women, and only 5 per cent of women farmers benefit from extension services. This, in combination with a continuing gap in access to resources, inputs, and technologies, negatively affects women farmers' ability to create sustainable livelihoods from their farms. (GFRAS, 2012: 2)

Closing the knowledge gap

So, government extension services are underfunded and in decline while donor or private sector-financed services are very limited in scale and tend to focus on specific (often cash) crops or are linked with finance packages for certain inputs (seed, fertilizer, herbicide, pesticide, etc.). Meanwhile, women, who play a major role in household food production, are largely ignored by all extension providers. As a result, smallholder farmers are finding it increasingly difficult to access the sort of technical knowledge that would help increase productivity on the marginal lands that most of them occupy.

But it is known that access to the right sort of technical knowledge can have a transformative effect on livelihoods and food security, as the example of floating gardens and sandbar pitting techniques from Bangladesh attests.

This is not an isolated example of the positive benefits of effective agricultural extension services. Studies to determine the returns on public investment in such services have demonstrated their value over a wide range of geographies. A review of a large-scale extension service programme in Uganda, for example, found that its benefit–cost ratio ranged from 1.3 to 2.7, translating to an internal rate of return (IRR) of 8 to 36 per cent (Tewodaj, 2012: 17). Similarly, a review of 81 extension programmes spanning a range of crops and livestock across Asia, Latin America, Africa, and a number of OECD countries reported that three-quarters of the programmes had IRRs greater than 20 per cent, and two-fifths with IRRs between 20 and 60 per cent (Tewodaj, 2012: 59).

Here is yet another example of a technology injustice: poor producers denied access to existing and potentially transformative technical knowledge as a result of widespread reductions in government expenditure (despite the positive returns on investment this particular spending accrues) and a tendency for private-sector services to cherry-pick the opportunities that offer the highest commercial return. Clearly, given the importance of agriculture to food security, rural prosperity, and wellbeing, a way must be found to reverse this trend and provide affordable advisory services to large numbers of small farmers on marginal lands with the aim of improving productivity through low-cost approaches involving minimal external inputs. There are many examples to draw on.

Some services seek to draw on new communications technology to expand the reach of the limited resources governments have. In Bangladesh, for example, the Ministry of Agriculture launched a free-phone service, the Krishi Call Centre, in June 2014 to provide advice to farmers on crops, livestock, and aquaculture, which was answering 10,000 enquiries a month by December that year (Bhuiyan, 2014; Practical Action, 2015). One hundred thousand farmers from two districts in Mashonaland Central Province, northern Zimbabwe, are now accessing advice from veterinary health and agricultural experts via instructions recorded in local languages on mp3 players (Nyathi, 2012), while in Peru the Infolactea website has around 40,000 hits a month from small dairy farmers accessing information encapsulated in simple YouTube videos on topics such as how to treat mastitis in cattle. While phone and internet-based services show clear potential to expand access to extension services, it is important to remember that they may not avoid some of the same problems faced by more traditional forms of agriculture extension, including a failure to reach out to women farmers. An evaluation of the Krishi Call Centre, for example, showed that just 2 per cent of those using the service were women, a fact the development partner agency, Practical Action, was still trying to understand the cause of at the time of writing.[2]

Other approaches seek to develop new forms of face-to-face, farmer-to-farmer extension services. In Cusco, Peru, a *Kamayoq* (the local name for agricultural extension workers) training school has been established to fill the gap in service provision to some of the poorest farming communities in the Andean mountains, particularly in the high-altitude areas over 4,000 metres, which depend almost entirely on alpacas and native potatoes. Despite the obvious need, these resource-poor farmers have had little access to any kind of technical support. Trainees are nominated by their own communities and include both women and men. Training is provided in Quechua, the local language. The course lasts eight months and involves attendance for one day per week. The emphasis is on practical learning, with training at different field locations focusing on local farmers' agricultural and animal health needs including: identification and treatment of pests and diseases of agricultural crops and livestock; better animal husbandry methods, particularly breeding; and irrigation. To date, 140 *Kamayoqs* have been trained. One of the objectives has been to encourage the *Kamayoqs* to work with farmers to generate their own solutions to local agricultural and veterinary problems, a process known as participatory technology development (PTD). This is important for two main reasons: firstly, active farmer participation builds farmers' confidence and is widely recognized as a key component of rural development. Secondly, the ability to innovate is itself important as conditions change and farmers need to be able to adapt (for example, to changing climatic conditions and new pests and diseases) (Hellin et al., 2006). Impacts have been positive, with mortality rates among cattle falling dramatically in the 30 communities among which the *Kamayoqs* were active and a pilot study in one community showing a 70 per cent increase in crop productivity. A good example of PTD in this process has been the discovery of a natural medicine to treat the parasitic disease liver fluke in alpaca. The *Kamayoqs* themselves have also benefited in terms of monthly income, ranging from an additional $88 for those focusing on cattle-feeding advice to $300 for those providing technical advice on irrigation (Practical Action, 2012).

Common to all these approaches is a belief that smallholder farming is critical to food security and rural prosperity, and that the decline in agricultural extension services has to be reversed to ensure the interests and needs of the many who work marginalized lands are not sacrificed in favour of the few who are able to farm commercially at scale on highly productive land.

The digital divide

The importance of ICTs for development

It may seem strange to include access to information and communication technologies (ICTs) as a critical element of establishing a basic social foundation, alongside access to basic services such as water, energy, food, and health. But as more and more services go online and access to a mobile phone or the internet becomes increasingly necessary to engage in everything from banking to filing a tax return, that's exactly what some national governments in the developed world are doing. Finland, for example, became the first country in the world to make access to broadband a legal right in 2010 (BBC, 2010) and, in 2015, the UK government announced that it too intended to ensure that access to fast broadband would be put 'on a similar footing as other basic services', giving everyone a legal right to request a 10 megabits per second connection (Department for Culture, Media & Sport, 2015).

But does the same hold true in the developing world? Can access to telephone services or the internet be said to be an essential element of a basic standard of living in countries such as Kenya or Nepal, in the same way it is in Finland and the UK? The answer to that question has to be a qualified 'yes'. 'Qualified' because the extent of the digital divide between those who have access to certain ICTs and those who do not is much larger in the developing world than in the developed and will be much more challenging to close as a result. But a qualified 'yes' because the use of ICTs, and particularly mobile telephony, is rapidly expanding in the developing world to host a whole range of essential services, from access to agricultural extension advice (discussed earlier) to attempts to bring citizens and government closer together.

Mobile phones also extend a fundamental communication service to many in an affordable manner for the first time, allowing people to maintain contact with family, friends, and more distant relatives, including partners or children who have moved from rural to urban areas to find work or further education. The contribution to people's wellbeing from this fundamental improvement in connectivity with friends and relatives should not be underestimated. To illustrate the additional benefits of improving access to ICTs, some examples of the broader potential developmental impacts are described briefly.

Mobile banking. Two and a half billion people in the developing world are 'unbanked' and have to rely on cash or informal financial services which are typically unsafe, inconvenient, and expensive. However, over 1 billion of these people have access to a mobile phone, which can provide a platform to extend the reach of financial services such as payments, transfers, insurance, savings, and credit (GSMA, 2014). At small shops and kiosks across the developing world people already buy vouchers to top up their mobile phone credit. Mobile money services take this a step further and enable these small traders to act as informal banks. They take cash and are able to credit it to your mobile money account via a special text message. Once your account is credited you can transfer funds, via another text message, to anyone else with a similar account and they can then withdraw the money by visiting a similar small trader in their own locality. Kenya represents the largest mobile money market with the total value of transactions made by mobile phone in 2013 representing more than half the country's GDP at $24 bn (*The Economist*, 2014). This is a rapidly growing industry, with mobile money services available in 61 per cent of all developing countries in 2014, through 2.2 million local service providers, meaning that in the developing world there are twice as many mobile money agents as cash machines and four times as many as local bank branches (GSMA, 2014). Sub-Saharan Africa leads the way with 62 million active mobile money accounts representing over half the world's mobile money activity (GSMA, 2014).

Mobile financial services are expanding beyond money transfers to include micro-insurance, for example, where the mobile money system is used to collect small regular premiums for life or medical insurance that would hitherto have been logistically too time-consuming and financially unprofitable to collect as cash. Mobile money systems are increasingly being seen as a means of driving 'financial inclusion', not just by providing banking facilities to those without bank accounts but also allowing marginalized and unbanked people to build up records of financial transactions that can provide evidence of creditworthiness and so open routes to loan finance.

Agricultural market information and trading. Mobile phone-based systems for providing market and technical information to farmers have expanded rapidly in recent years. These include services such as MFarm in Kenya, which started by offering small farmers information on the latest market prices for their crops via an SMS messaging system, and now has a service to link farmers together to enable them to pool their harvests and get better prices from dealers by trading in larger volumes (Solon, 2013). Other large commercial providers are also moving into the market. Reuters Market Light (RML) has been providing mobile services to small farmers in India since 2007 and was spun off as a separate company by Thomson Reuters in 2013. Farmers receive around five SMS texts each day containing information to help with planning when to plant and harvest, how to improve cultivation and control pests and diseases, as well as advice on how to access government subsidies and negotiate better prices for their crops. RML now has more than 1.4 million smallholder farmer subscribers across 17 Indian states (Vodaphone, 2015). New variants of mobile services for farmers are being developed all the time. For example, Tech for Trade, a UK-based non-government organization and investor in MFarm, has introduced a new trading platform that seeks to build greater trust between small farmers, buyers, and traders by introducing transparency throughout deals. Each stage of a trading deal is recorded on an online platform which is accessible from a mobile, tablet, or computer, allowing everyone involved, including farmers and end buyers, to know exactly what costs are incurred by the 'middle man' and what prices are being offered at every stage in the transaction (Tech for Trade, 2015).

Remote health services. One of the challenges of providing health-care in relatively remote rural locations is the difficulty of recruiting doctors willing to live and work in such places. Moreover, the cost of providing specialist care (orthopaedics, ear, nose, and throat, paediatrics, etc.) in remote locations is considered prohibitive by many health services. Rural patients often have to travel long distances to district or regional headquarters to find anything beyond diagnosis of the most basic kind. This can result in delayed or deferred treatment, impacting on health, or additional costs and lost income as a result of time spent travelling. Remote health diagnostics that allow patients in rural locations to be seen via a webcam by specialists in distant hospitals offers one solution to this problem. The Aravind Eye Hospital in the southern Indian state of Tamil Nadu is a good example. In 2004, using a wireless broadband network, the hospital was able to connect up five rural clinics so as to provide eye services to thousands of rural residents. Aravind doctors were able to screen about 1,500 patients each month via a videoconferencing system and so allow patients with minor eye problems to be diagnosed and resolved locally, reducing the numbers who had to travel long distances to the hospital to those with more serious conditions. The system was successful enough to be later scaled up (World Bank, 2009). The provision of remote health services is growing fast and now includes a range of services, from support for maternal and newborn health in Mongolia and breast screening in Mexico (WHO, 2010), to a mobile phone app in Myanmar that can help a patient with hepatitis B find a certified doctor, clean needles and antibiotics, read health information about the management of their condition, and order high-quality medication for treatment (Bosier, 2015).

Digital identity. In many developing countries, significant numbers of people remain without official identification. The World Bank reported that in sub-Saharan Africa, as many as 55 per cent of the population had no official identification record, while globally 625 million children were not registered at birth, more than 2 billion people lacked formal identification, and 500 million people were outside the regulated financial system as a result of lacking recognized identity documentation (World Bank, 2015). Digital identity – the issuing of a unique and secure identification and authentication of a person's identity – is seen by many as an important way to bridge this gap, particularly as it can allow poor people access to important government safety net programmes and subsidies. Attempts to address this by governments in the developing world have ranged from the basic digitization of paper-based birth certificates in Peru, to make them more easily accessible online (Chaudhary, 2014), to the much more ambitious and somewhat controversial Aadhaar programme of the government of India. Started in 2010, the Aadhaar programme aimed to capture the fingerprints and iris scans of all 1.2 billion Indians in order to assign each a 12-digit digital identity that could be verified instantly online, providing a secure and quick way of verifying identify for everything from access to kerosene subsidies to opening a bank account. About three-quarters of the population (920 million) had been registered by late 2015 (Kumar, 2015). The concerns of civil-society groups around privacy issues and security of data led the Indian Supreme Court to rule in March 2014 that the national government could not make Aadhaar IDs mandatory for receiving government subsidies and that the department which manages the Aadhaar programme should not share any personal information with any third party without the permission of the individual.

Citizen action and voice. SMS messaging groups have proven popular in helping people organize themselves and share information, sometimes in response to a lack of transparency on behalf of governments. FrontlineSMS is one of the well known, a generic communications platform developed by innovator Ken Banks that can be downloaded free from the internet. To use FrontlineSMS requires just a laptop computer and a mobile. After downloading the software and attaching a mobile phone to the laptop with a cable, users can create contact lists of people and send SMS messages to all simultaneously. The system has been used by farmers in Indonesia, Cambodia, Niger, and El Salvador to exchange information on up-to-date market prices for their produce, and by civil-society groups in Zimbabwe and Pakistan to set up two-way messaging without needing to go through local operators to keep citizens in disconnected rural areas informed of events during periods of political upheaval (National Geographic, 2015). Ushahidi is another well-known non-profit that has developed open-source software and a web platform to help people in any part of the world to collect and broadcast information about a crisis. Individuals can provide information by text message (sometimes in combination with FrontlineSMS), email, or web postings, and the software aggregates and organizes the data into a map or timeline. A new product called Swift River goes further and uses machine-learning algorithms to automatically extract and organize information from text messages, emails, blogs, and tweets. Ushahidi's crisis-mapping software was first used to track outbreaks of violence in the aftermath of the disputed 2007 Kenyan general election and has subsequently been used to track and coordinate such events as disaster relief following the Haiti earthquake in January 2010, to a snow clean-up in New York City (MIT Technology Review, 2015).

These examples show some of the potential for ICTs to have a positive (if sometimes controversial) impact on the lives of the poor in the developing world. Studies by the International Development Research Centre (IDRC) seem to confirm this, finding links between ICT access and reduced poverty among the very poor. A three-year study, for example, compared a large cohort of Peruvians who had become internet users to a group of non-users. At the end of the study the household incomes of the internet users were on average 19 per cent higher than the non-users. Likewise, in a study of two villages in Tanzania, residents of one village received five months of mobile phone airtime and internet access, while residents of the other village did not. The first village experienced a reduction across all seven of the indicators used to measure poverty in the study, while in the control village improvements were seen in only two of the seven indicators (Elder et al., 2013). Beyond these individual examples, more generalized evidence of positive correlation between access to ICTs and economic growth is emerging. The World Bank's

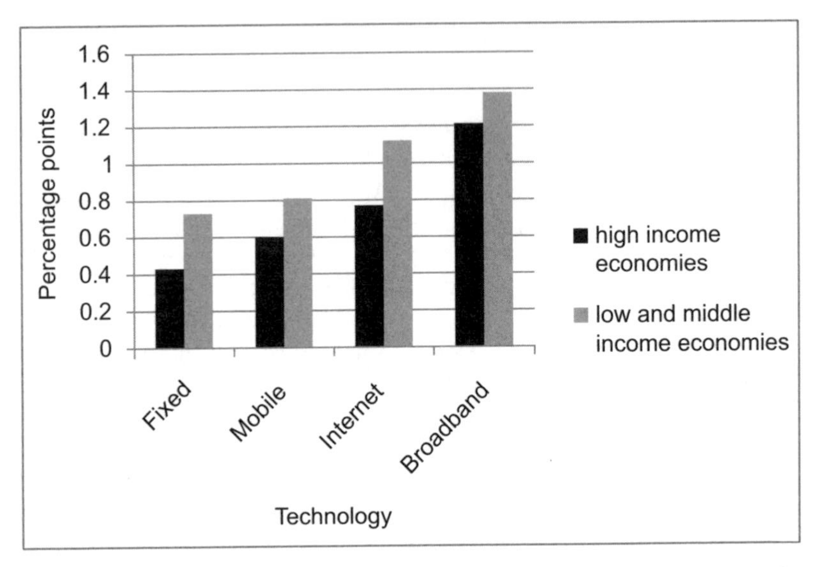

Figure 4.2 Percentage increase in economic growth per 10 per cent increase in penetration of telecommunication technologies

Source: World Bank, 2009

review of economic impact assessments in 2009, for instance, showed the growth benefit that broadband provides for developing countries was about a 1.38 percentage point increase for each 10 per cent increase in penetration (see Figure 4.2). Considering all the above, it would seem that access to ICTs has the potential to make a positive contribution to all 11 elements of Raworth's social foundation.

Digital inequalities

In many ways the rapid expansion and take-up of mobile phone services in Africa and Asia is a success in stark contrast to the rate of expansion of other services on the two continents. The number of people with access to a mobile in India had already exceeded the number with access to a toilet by 2010 (United Nations University, 2010). In addition, mobile phone coverage is the most widely available infrastructure across rural and urban Africa, ahead of electricity, piped water, roads, and sewerage services (Mungai, 2015). However, there are still notable divides.

Developed/developing nation divide: Although mobile cellular subscriptions in the least developed nations have risen dramatically from single figures in 2005 to around 59 per cent of the population by 2013, that is still well below the global average of 96 per cent of the population and the developed world average of 126 per cent. Moreover, it is known that mobile subscription figures far exceed the actual number of mobile phone users and that a more accurate view of the number of people with mobile phones in the least developed countries could be as low as 30 per cent of the population (ITU, 2014).

Urban/rural divide: As is the case with water and electricity services, there is a marked difference between urban and rural access rates for mobile phone services in many developing countries, often as a result of the high cost of extending masts and infrastructure into remote rural areas combined with the absence or unreliability of electricity. In Africa, 129 million people still live in areas not covered by mobile phone signals, while 309 million people are in that situation in Asia. In India, 82 per cent of urban households have access to a telephone compared with just 54 per cent of rural households (ITU, 2014).

Gender divide: In low to middle-income countries a woman is 21 per cent less likely to own a mobile phone than a man, with this figure increasing to 23 per cent if she lives in sub-Saharan Africa, 24 per cent in the Middle East, and 37 per cent if she lives in South Asia (GSMA, 2010). Reasons cited in the GSMA survey for these discrepancies were:

1. The total cost of ownership of a mobile phone. The report noted that women are often financially dependent on men or do not have control over economic resources, which makes accessing ICT services more difficult, particularly where access to credit is gained through male household members.
2. Cultural barriers to mobile phone ownership and access to ICTs by women. In some societies, women are barred from public places, making access to public calling offices, community telecentres, or internet kiosks difficult for them.
3. Limited technical literacy among women at the base of the pyramid. Seventy-five per cent of women with monthly household incomes below $75 do not own a mobile phone. Of that group, more than a third 'expressed concerns about the complexity of the technology or the level of literacy required to use a mobile device' (GSMA, 2010: 3). Allocation of resources for education and training often favours boys and men, resulting in lower levels of literacy and education, including training in languages which are predominantly used in ICT platforms and the internet (GSMA, 2010).

When you look at access to broadband and internet usage the gap between the developed and developing nations looks significantly wider. The penetration of fixed-line broadband remains low in most developing countries due to the expense of the cabling and infrastructure required. Mobile broadband is, by contrast, expanding more rapidly and seems to be the preferred choice of many. Developing countries are still largely reliant on 2G networks, however, and broadband data plans and smartphone costs remain high, meaning fast mobile broadband is generally confined to urban centres and better-off consumers. As a result, the penetration of mobile broadband in the least developed countries was just 12.1 per cent in 2015, compared to 86.7 per cent in the developed world (UNESCO, 2015).

Fast growth but a long way still to go

Mobile telephony is often referred to as a technology success story in the developing world – an example of technology transfer, adaptation, and consumer-led demand resulting in an explosion of new services and opportunities, such as mobile money, remote health services, agricultural advice lines, and crowd-sourcing communications platforms, many of which have a direct relevance to the lives of the poorest in society. While this is true to an extent, and makes mobile telephony arguably the most innovative technological sector in the developing world, much remains to be done to ensure universal access to affordable mobile and broadband services in rural as well as urban locations.

The digital divide is not just about access to affordable infrastructure either, but also about issues of power and access to resources within households that reveal themselves in the discrepancy of access rates between men and women. In the final analysis, access to ICTs is fundamentally about access to knowledge, and so the ability to benefit from the former will, inevitably, accrue most readily to those who, once connected, are best able to navigate and exploit the latter. An interesting quote from a lecture given by Spanish academic Manuel Castells sums this up well, noting that as more and more people are gaining access to mobiles and other forms of ICT:

> connectivity as an element of social divide is quickly losing relevance. But what is in fact observed in those persons that are connected, particularly students and children, is that a second element of social divide appears, much more important than technical connectivity, that is the social and cultural capability for using [the] Internet. Once all that information is in the web ... the coded knowledge, but not the tacit knowledge that is needed for what one wants to do, the issue is to know where the information is, how to search, process and transform it into specific knowledge ... That capability of learning to learn, to know what to do with what one learns, that capability is socially unequal and is related to social and family background, to cultural and educational level. That is the place, empirically talking, of the digital divide at this moment ... (Arocena and Sutz, 2001)

Illiteracy therefore becomes an even greater cause of social and economic divide, excluding over 770 million adults worldwide from largely text-based digital services and forcing those unable to read to communicate on mobile by voice rather than by cheaper SMS services. Furthermore, the lack of local language content and the predominance of material in English on mobile internet can also diminish the benefits of access even for those who can read (GSMA, 2014).

Access to ICTs may not *yet* be a technology injustice, in that it is actually an example of a new technology being rapidly deployed with genuine positive impacts for those less well off. However, inequalities of capacity to access digital services between rural and urban populations, men and women, and the literate and illiterate, combined with a lack of digital content in local languages, could lead to a widening digital divide of the nature referred to by Castell and exacerbate existing social inequalities. The challenge will be to maintain the momentum of extending mobile services to those hard-to-reach populations to prevent this happening. Suggestions on how to achieve this, from the IDRC study referred to earlier, include:

- opening telecommunications markets to competition so that prices come down. Ethiopia is cited as an example of where the state's retention of a monopoly keeps prices high and penetration low. Issuing new licences to telecommunications players is the most readily available intervention;
- supporting the ICT sector with government funding to help with infrastructure costs;
- reducing taxes on communication services, which should be seen as vital for society, rather than a luxury;
- nurturing the creation of useful content and promoting decentralized innovation by supporting incubators, fostering interactions among entrepreneurs, and encouraging investors;
- offering training in basic computer literacy and the skills needed to fulfil job requirements (Elder et al., 2013).

Solutions proposed for the expansion of the rural mast networks necessary for better coverage include infrastructure sharing. At one level this requires regulation to encourage mobile service providers to share masts and avoid costly duplication, but it can also mean including the power requirements of rural masts as a guaranteed base load with the potential to make rural (renewable energy) electricity mini-grids more affordable, thus helping to extend both rural electricity and mobile services at the same time (GSMA, 2014).

Technology Justice and the social foundation

Chapters 3 and 4 of this book have looked at examples of injustices in access to technology under 4 of the 11 constituents of Kate Raworth's social foundation: energy, water, food, and health, plus a fifth cross-cutting technology area, access to ICTs. It would be possible to examine technology injustices around access in the remaining seven areas (education, income, gender equality, social equity, voice, jobs, and resilience) although, as Table 2.3 in Chapter 2 suggests, many of these would, in turn, most likely have origins in technology injustices under the five areas of technology access already reviewed.

In each of the cases examined it is clear that an injustice arises because:

- access to certain technologies or technical knowledge is critical to attaining a basic minimum standard of living (Raworth's social foundation), but
- although the necessary technology or technical knowledge already exists and may be in use in the developed world, it remains out of reach of the poor for reasons including a lack of prioritization in public financing (for example, energy services); a lack of 'expressed demand' (affordability by any other name) where responsibility for provision has been delegated to the market (for example, privatized agricultural extension services); a clash of interest with patents or intellectual property rights (for example, HIV/AIDS retroviral drugs); or even a lack of land tenure rights (for example, urban water and sanitation services).

Removing these barriers requires, first and foremost, recognition in public policy of the critical role access to technology plays in underpinning a basic standard of living. This recognition has not always been present in recent years. The Millennium Development Goals (MDGs), for example, had specific targets for water and sanitation coverage which resulted in a degree of public attention being paid to global and national progress on these issues. On the other hand, the MDGs did not refer to the need for access to basic energy services. That omission resulted in a 15-year gap in both sustained creative thinking about the problem and pressure from global institutions on national governments to address it. The United Nations' Sustainable Energy for All (SE4All) initiative and the incorporation of energy access into the Sustainable Development Goals (SDGs) has already had a dramatic impact on the level of attention drawn to the issue, reinforcing the idea that global public policy initiatives are at least part of the solution. Indeed, as noted in the introduction to this book, the explicit recognition of the importance of technology across both the SDG and climate change processes offers hope that, at least at an international level, public policymakers now see how critical technology access is to creating a basic universal standard of living. The potential of the SDGs' Technology Facilitation Mechanism and Technology Bank and the Technology Mechanism proposal that resulted from the climate change talks are discussed more in the following chapters, but it is probably too early to assess their impact or, in the case of the SDG mechanisms, even their potential impact.

Beyond the rhetoric of the SDGs, it is also important to recognize that a serious attempt to address the technology injustices of access will require a radical rethink of widely accepted policy narratives. This in itself will be a major challenge as, at best, it will require running up against bureaucratic inertia and, at worst, clashing with large vested interests. As noted earlier, 65 per cent of the additional investment necessary to provide universal electricity services by 2030 will need to be in off-grid technologies. But the energy sector and those who finance it are structured to build large power stations and extensions to national grids. They are neither minded to, nor necessarily capable of, delivering solar home systems, improved cook-stoves, and micro-generation projects. The water sector faces similar problems. In the health sector, universal access to essential medicines will only be achieved if major pharmaceutical companies can be persuaded to adopt differential pricing strategies for those drugs. Meanwhile, ministries of agriculture have to rethink their assumption that prioritizing support to commercial farmers on prime land over smallholder farmers on marginal lands will deliver food security. They further need to accept that neither the traditional and inefficient government extension services, nor the new breed of commercially financed advice, will deliver the transformation needed in the fortunes of the rural poor.

In this respect, achieving Technology Justice is fundamentally a political project. Although access to essential technologies is the desired outcome, this will only happen if poor people's voices are heard in national planning processes and if their expression of need leads to a reprioritizing of public policy objectives. This is indeed a tall order but not an impossible one. International initiatives can provide some momentum for a rearticulation of policy priorities from the perspective of the poor. The Global Tracking Framework of the SE4All initiative is a good example of such a refocusing of a policy narrative, in this case from defining progress on energy access in terms of 'grid connections' to a definition based on whether people have access to the range of basic energy services they need, regardless of whether that energy is derived from grid or off-grid sources.

Global initiatives to reframe policy narratives alone are not enough and seldom have the power or the inclination to really challenge entrenched interests. Without strong domestic pressure for change it is unlikely, for example, that the SE4All initiative will manage to persuade governments to adopt the multitier Global Tracking Framework for their national statistics on energy access, especially given that its use is likely to result in lower (but more realistic) levels of access being recorded than have been the case in the past – something that would be politically difficult to swallow for many countries. Real change therefore requires a strong civil society and an aware and vocal national constituency of active citizens that can exert the relentless political pressure required.

Finally, something that has not yet been addressed in this book, is the importance of recognizing the interdependence of many of the basic services required for the social foundation. There is a clear nexus between the provision of food and access to basic water and energy services, for example (see Figure 4.3).

Agriculture requires water to grow crops and energy to power irrigation, for traction, crop drying and storage, and the production of fertilizer and other inputs. In return, agriculture can produce fuel from agricultural waste or the growing of biofuel crops to make energy, and cropping patterns can have beneficial effects on a water catchment's absorption of rainfall and storage of groundwater. Likewise, energy production often requires water for steam generation, power-station cooling, or hydro-electricity generation, while energy

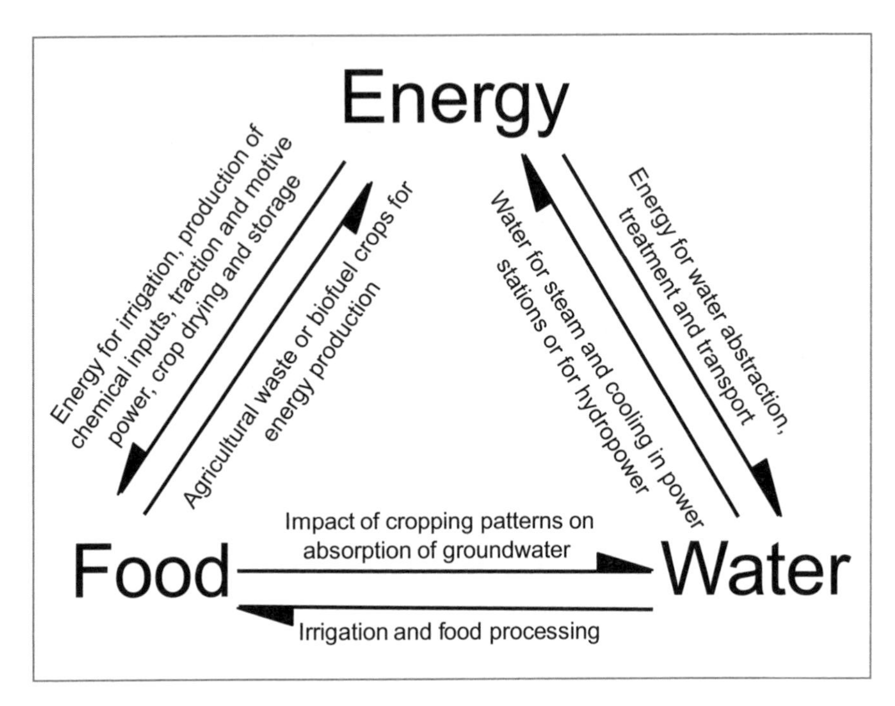

Figure 4.3 The energy–food–water nexus

is required for water extraction, treatment, and transport. This nexus highlights the need to consider the trade-offs that occur as a result of different approaches to the delivery of basic services which, in turn, requires a more systemic rather than sectoral approach to national and regional planning. This constitutes another significant change of approach for policymakers and is a theme that will be returned to when considering innovation systems in the following chapters.

Notes

1. The Gaibandha example is drawn from the work of Practical Action and the text is based on the author's field trip notes and personal correspondence with Practical Action's Bangladesh office staff.
2. Women's limited daytime access to phones is thought to be at least part of the explanation. Men generally had possession and control of the household phone and were typically away from the home during the time the service was operational each day (office hours), preventing women from access.

References

ActionAid (2013) *Walking the Talk: Why and How African Governments Should Transform their Agriculture Spending*, London: ActionAid.

AGRA (2014) *Africa Agriculture Status Report 2014*, Nairobi: Alliance for a Green Revolution in Africa.

Arocena, R. and Sutz, J. (2001) 'Innovation systems and developing countries', Working Paper no. 02-05, Copenhagen: Danish Research Unit for Industrial Dynamics <http://www3.druid.dk/wp/20020005.pdf>.

BBC (2010) 'Finland makes broadband a "legal right"', *BBC News* [online], 1 July <http://www.bbc.co.uk/news/10461048> [accessed 19 November 2015].

Bhatkal, T., Samman, E., and Stuart, E. (2015) 'Leave no one behind: the real bottom billion', London: Overseas Development Institute.

Bhuiyan, M.K.I. (2014a) 'Journey of Krishi Call Centre: Hello......16123', *Practical Action Blog*, 19 August <http://practicalaction.org/blog/news/journey-of-krishi-call-centre-hello16123/> [accessed 13 July 2015].

Bientema, N. and Fuglie, K. (2012) *ASTI Global Assessment of Agricultural R&D Spending*, Washington DC: International Food Policy Research Institute.

Bosier, C. (2015) 'How smartphones fill the healthcare gap in Myanmar', *National Geographic* [online], 22 September <http://voices.nationalgeographic.com/2015/09/22/how-smartphones-fill-the-healthcare-gap-in-myanmar/> [accessed 19 November 2015].

Chaudhary, S. (2014) 'What's in a name?', *Practical Action Blog*, 18 December <http://practicalaction.org/blog/news/whats-in-a-name/> [accessed 20 November 2015].

De Schutter, O. (2009) 'The right to food and the political economy of hunger', 18th McDougall Memorial Lecture, Rome, 29 November.

Department for Culture, Media & Sport (2015) 'Government plans to make sure no-one is left behind on broadband access', *Gov.uk*, 7 November, <https://www.gov.uk/government/news/government-plans-to-make-sure-no-one-is-left-behind-on-broadband-access> [accessed 19 November 2015].

Elder, L., Rohan, S., Gillwald, A., and Hernan, G. (2013) *Information Lives of the Poor: Fighting Poverty with Technology*, Ottawa, Canada: International Development Research Centre.

ETC Group (2009) 'Who will feed us? Questions for the food and climate crises', Communiqué 102, Ottawa: ETC Group.

FAO (2006) *The State of Food Insecurity in the World 2006: Eradicating World Hunger – Taking Stock Ten Years after the World Food Summit*, Rome: Food and Agriculture Organization of the United Nations.

FAO (2009) *The State of Food Insecurity in the World 2009: Economic Crises – Impacts and Lessons Learned*, Rome: FAO.

GFRAS (2012) 'Fact sheet on extension services position paper', Lindau: Global Forum for Rural Advisory Services.

GSMA (2010) 'GSMA mWomen Programme: Policy recommendations to address the mobile phone gender gap', London: GSMA.

GSMA (2014) *Digital Inclusion 2014*, London: GSMA.

Hellin, J., De la Torre, C., Coello, J., and Rodriguez, D. (2006) 'The Kamayoq in Peru: farmer-to-farmer extension and experimentation', *Leisa Magazine*, 22(3): 32–34.

ITU (2014) *Measuring the Information Society Report*, Geneva: International Telecommunications Union.

Infolactea (no date) <http://infolactea.com/>.

Kumar, S. (2015) 'Modi govt goes all out for Aadhaar', *Live Mint*, 6 October, <http://www.livemint.com/Politics/9n945jW14A93TRD2YfHuKJ/NDA-govt-goes-all-out-for-Aadhaar.html> [20 November 2015].

MIT Technology Review (2015) 'Ushahidi', <http://www2.technologyreview.com/tr50/ushahidi/> [20 November 2015].

Mungai, C. (2015) 'The mobile phone comes first in Africa; before electricity, water, toilets or even food', *Mail and Guardian Africa*, 3 March, <http://mgafrica.com/article/2015-03-03-mobile-come-first-in-africa-before-electricity-water-toilets-or-even-food> [20 November 2015].

National Geographic (2015) 'Explorers Bio: Ken Banks – mobile technology innovator', <http://www.nationalgeographic.com/explorers/bios/ken-banks/> [accessed 20 November 2015].

Nyathi, T. (2012) 'Local language podcasting in Zimbabwe', *New Agriculturist*, <http://www.new-ag.info/en/focus/focusItem.php?a=2486> [accessed 13 July 2015].

OECD (2006) *Promoting Pro-Poor Growth: Agriculture*. Paris: Organisation for Economic Co-operation and Development.

Practical Action (2012) 'The new Kamayoq – Peru', <http://practicalaction.org/the-new-kamayoq-peru> [accessed 13 July 2015].

Practical Action (2015) 'Krishi Call Centre nominated for WSIS global award', 20 March, <http://practicalaction.org/krishi-call-centre-wsis-nomination> [accessed 13 July 2015].

Shakhovskoy, M. (2014) 'Technical assistance for smallholder farmers: an anatomy of the market', Briefing 07, The Initiative for Smallholder Finance.

Solon, O. (2013) 'MFarm empowers Kenya's farmers with price transparency and market access', *Wired*, 21 June <http://www.wired.co.uk/news/archive/2013-06/21/mfarm> [accessed 20 November 2015].

Tech for Trade (2015) 'Trade transparency', 18 September <http://tt.techfortrade.org/en/> [accessed 20 November 2015].

Tewodaj, M. (2012) 'The impacts of public investment in and for agriculture: synthesis of the existing evidence', Rome: FAO.

The Economist (2014) 'Mobile money in developing countries', 20 September <http://www.economist.com/news/economic-and-financial-indicators/21618842-mobile-money-developing-countries> [accessed 19 November 2015].

UNESCO (2015) *The State of Broadband*, Geneva: UNESCO and ITU.

United Nations University (2010) 'Greater access to cell phones than toilets in India: UN', 14 April <http://unu.edu/media-relations/releases/greater-access-to-cell-phones-than-toilets-in-india.html> [accessed 20 November 2015].

Vodaphone (2015) 'Connected farming in India: how mobile can support farmers' livelihoods', Newbury, UK: Vodaphone Foundation.

WHO (2010) *Telemedicine: Opportunities and Developments in Member States*, Geneva: World Health Organization.

World Bank (2007) *World Development Report 2008: Agriculture for Development*, Washington, DC: World Bank.

World Bank (2009) *Information and Communications for Development 2009: Extending Reach and Increasing Impact*, Washington, DC: World Bank.

World Bank (2015) 'Digital IDs for development', 6 May, <http://www.worldbank.org/en/topic/ict/brief/digital-ids-for-development> [accessed 20 November 2015].

CHAPTER 5
Technology Justice and use

Chapters 3 and 4 used the lens of justice as fairness to look at people's access to technologies critical to the attainment of a basic standard of living – Raworth's social foundation. This chapter moves on to use the concept of justice as compromise to explore how the use of technology by a person or persons may impinge on the ability of others now, or in the future, to live the lives they value and attain for themselves a basic standard of living.

Justice as compromise

It is possible to find examples of injustices in technology use that relate to Raworth's social foundation. In the 1980s clean drinking water was provided in the Barind area of Rajshai Division in northwest Bangladesh through the installation of numerous tube wells fitted with handpumps. These provided freshwater supplies for rural households for many years. By the mid-1990s, however, wealthier farmers had started to use motorized pumps to extract large amounts of groundwater for dry-season irrigation of crops, to the extent that in the dry season the water table (the level beneath the ground's surface at which, if you dig a hole, you encounter water) had started to drop. This lowering of the water table put groundwater out of reach of most of the suction handpumps[1] that had been installed for domestic water supply. An injustice arose from some farmers using motorized pumps for irrigation as this resulted in less well-off families losing their safe drinking-water supply for several months each year.[2]

A greater injustice in technology use that relates to Raworth's social foundation is the relationship that is often referred to as 'fuel versus food'. The impact of biofuels on global food prices has been the subject of considerable discussion, debate, and media attention since 2007 (EBTP, 2012). In 2008, European Union member states committed to obtaining 10 per cent of transport fuels from renewable sources by 2020. This was, in effect, a commitment to a massive increase in the production of biofuels. The question here is whether, as a consequence of grain now being traded as a commodity on international markets, Europe's (and, for that matter, the US's) technology choice of biofuels to fuel its cars and trucks has had an adverse impact on the price of staple foods in the developing world.

The non-government organization ActionAid produced a report in 2010 arguing that the diversion of food crops into the production of biofuels (for example, corn into ethanol) had indeed resulted in increased scarcity of food grains on international markets, which was, in turn, responsible 'for at

http://dx.doi.org/10.3362/9781780449043.006

least 30 per cent of the global food price spike in 2008 ... [which] pushed ... about 30 million more people into hunger' (ActionAid, 2010). Two years later ActionAid produced a second report looking at the impact of corn production for ethanol in the US, noting that the percentage of the US crop which was diverted into ethanol production had increased from 5 per cent in 2000 to 40 per cent in 2012. ActionAid linked this to a rise in the cost of maize on international markets over the same timeframe and a 70 per cent increase in the price of the maize-based staple food *tortillas* in Mexico over the period 2005–11 (ActionAid, 2012).

Although there have been many claims and counter-claims as to whether there is a link between food prices and diversion of crops to biofuels, several studies beyond ActionAid's work suggest that this is so. The University of California's Giannini Foundation of Agricultural Economics, for example, concluded that, based on 2007 data, 'ethanol raised corn prices at least 18 per cent and perhaps as much as 39 per cent, depending on elasticity assumptions' (Wharton University, 2013). Concerns have also been raised that the emphasis on biofuels in the EU is leading to massive land grabs in Africa for the industrial-scale production of biofuel crops, often displacing smallholder farmers from their lands in the process (ActionAid, 2010).

There are many illustrations of how the use of technology by one group of people has a negative impact on another group's ability to achieve the standard of living summarized in Raworth's minimum social foundation. There are also many compelling examples of injustices in the use of technology that relate to breaches of Rockström's planetary boundaries and the impact this will have on future generations. This chapter begins by considering in more depth two examples of the latter (the impact of today's industrialized agriculture on future food security and the effect our use of fossil fuel technologies will have on the future climate) before going on to examine one of the former (the risk to the health of future generations from our use of antibiotics today).

Industrialized agriculture and biodiversity loss

The problem with large-scale commercial farming extends beyond the monopolization of resources available for extension and research discussed in Chapter 4. Industrialized agriculture itself has many problems, particularly the loss of genetic diversity within the species of livestock and crops we rely on for food. This narrowing of the genetic base of our food chain poses a significant risk to the long-term security of global food supplies.

It is helpful to start by looking at a specific case from Sri Lanka that illustrates both the value of genetic diversity in agriculture and the inability of the sector's formal institutes to conserve it. One of the impacts of the 2004 tsunami in Sri Lanka was an increase in salinity levels in the soil in many of the eastern coastal areas as a result of the inundation by the sea. This exacerbated a longstanding problem with salinization of coastal soils in areas such as those adjoining the estuary of the Walawe River in Hambanthota District, on the southeastern corner of the island. Over the past 20 years, rice farming, a staple source of food and income for smallholder farmers, has become increasingly difficult in this area because of the low tolerance of commonly cultivated varieties of paddy rice to salt. Yields have been dropping and many paddy fields have been abandoned as unprofitable. Although some farmers on the less badly affected land had some success with saline-resistant varieties developed at the local Ambalantota Rice Research Institute, in the worse affected areas even the resistant varieties were failing (Berger, 2009).

One attempt to address the problem was to go back to traditional varieties of rice and traditional organic cultivation techniques to improve the soil condition, as described by one farmer involved in the trials:[3]

> I am Ranjith. I took up paddy farming just like my forefathers before me. Our paddy lands have always had a high level of salinity due to the proximity to the sea and harvests have been low. The sea water that came in to our paddy with the Tsunami of 2004 further aggravated this condition. Due to the high level of salinity in the field, paddy seedlings started dying. Little by little, with each season (there are two planting seasons for rice each year), the harvest reduced. After the third season it became almost impossible to plant paddy. The modern varieties of paddy we were used to growing were unsuccessful in this highly saline condition. We were on the verge of abandoning the only form of livelihood we knew. It was at this crucial juncture (September 2005) that two organizations, namely Practical Action and the National Federation for the Conservation of Traditional Seeds and Agricultural Resources (NFCTSAR), came forward to help us. NFCTSAR identified 10 traditional rice varieties from its previous work that were suitable for saline soils and suggested we cultivated them on a trial basis. NFCTSAR provided us the required seed paddy. They also trained us on appropriate cultivation methods. Sixteen farmers in this area (including myself) tried out these traditional varieties for three seasons. At first we were rather sceptical. However, to our surprise and delight, seven of the varieties did in fact flourish in the saline conditions. We used organic fertilizer instead of chemical fertilizers for growing these traditional varieties, as recommended by the organizations. During the same period, a modern hybrid paddy variety was cultivated in the paddy field adjoining mine. This paddy field was fertilized with costly chemical fertilizer. Pesticides had to be sprayed as well to safeguard the crop from pest attacks. Finally, this was largely unsuccessful. I, on the other hand, used only organic fertilizer, the raw material for which could be easily sourced within the village with minimal cost. Pesticides were unnecessary since the indigenous seed paddy was capable of resisting pest attacks. I realized that if I grew these varieties commercially, the cost of production could be reduced significantly. Our trials revealed another unexpected result. In the case of certain saline-resistant traditional rice varieties, such as *Rathdel, Dahanala, Madathawalu,* and *Pachchaperumal*, the yields were

> high. Earlier, when we grew modern paddy varieties, we got only 15 to 20 bushels from an acre. Now with these traditional indigenous varieties of paddy, yields can be as high as 60 to 70 bushels per acre.

It should be noted that in this case the 'solution' not only tackled the salinization problem, but also increased yields and produced a product which could be labelled as 'organic' and so fetched a premium price at markets in Colombo.

This case study from Sri Lanka, while specific to a particular location and time, raises three interesting and more general points:

1. The ability to fall back on indigenous knowledge proved to be a valid and valuable solution to what appeared to be an increasingly unsustainable form of food production. Traditional cultivation techniques, which involved composting, the use of locally prepared organic liquid fertilizer, and, occasionally, pesticides prepared from the leaf of the neem tree, improved the soil structure and fertility, and controlled pests. The availability of a wide range of traditional varieties of rice meant 10 alternatives with known salt tolerance could be further tested to find a subset of varieties that worked best in that particular location.
2. This problem is not an isolated one but something that is expected to become increasingly common with predicted climate change. Storm surges and sea-level rises are likely to increase levels of salinity in coastal soils not only in Sri Lanka, but in many other locations as well.
3. The ability of formal agricultural sector institutions to provide a solution in this case was very weak. The Sri Lankan government's Rice Research and Development Institute (RRDI) had developed a very limited number of salt-tolerant varieties of paddy, none of which were up to the challenge. Instead, farmers had to rely on the work of NFCTSAR, a small farmer-led local NGO committed to maintaining traditional crop varieties and cultivation techniques. It turned out that the government agricultural extension service, which was used to meeting the needs of larger commercial farmers using modern seeds, fertilizers, and agrochemicals, was also not well equipped to provide advice on organic cultivation techniques. Nor was it able, for that matter, to offer advice to others in the market chain, such as rice mill operators requiring information on changes to their hygiene practices in order to mill rice labelled as organic (and who proved very willing to learn).

The links between biodiversity and long-term food security

The value of biodiversity in our food systems goes well beyond the provision of salt-tolerant rice varieties. What is more, the lack of engagement of the Sri Lankan RRDI in the maintenance of traditional varieties to support that biodiversity is a symptom of a much bigger flaw in food production systems. To understand this, it is necessary to grasp how the industrialization of food production under the green revolution is inextricably bound to the narrowing of the genetic base of our food chain.

The ETC Group, writing in Paul Pojman's book *Food Ethics*, notes that before the threat of climate change was recognized, government efforts to conserve plant and livestock species (through seed and sperm banks, for example) focused on the characteristics of yield and uniformity to meet industrial processing requirements and support profitability. As the understanding of the potential impact of climate change has grown and as the global food system has undergone other stresses, such as the dramatic spike in global food prices during 2007 and 2008, it has become apparent that the food chain is under threat. It is now necessary to think about genetic diversity in livestock and crops as a way of mitigating risk.

Industrial farming has displaced local varieties of crops:

> The Green Revolution's focus on wheat, rice and maize and commercial breeders' focus on soybeans, alfalfa, cotton and oilseed rape has pushed other traditional food crops into the margins since the 1960s. But the focus on yield has also meant that even within the world's leading crops it is estimated that genetic diversity has been decreasing by 2 per cent per annum since the 1990s and that perhaps three quarters of the germplasm pool for these crops is already extinct. (ETC Group, 2012: 186)

This severely limits the genetic pool that can be drawn on to develop crops that can cope with new climatic conditions, pests, and diseases in the future.

The same issue is faced in the livestock sector where the search for uniformity and productivity has led to a focus on a very narrow range of breeds globally:

> On average just five breeds dominate commercial production in each of the five main species of livestock around the world. Holstein-Friesian dairy cows are found in 128 countries for example, whilst the Large White pig is farmed in 117 countries, and the White Leghorn Chicken is found almost everywhere. (ETC Group, 2012: 186)

Although expanded farming of these five breeds has resulted in increases in productivity, measured by the amount of meat produced per animal, for example, this narrowing of the gene pool carries with it real risk: Avian influenza and Mexican 'swine flu' (H1N1) are just two recent examples of global pandemics largely provoked by extreme genetic uniformity in commercial breeds raised in confined spaces (ETC Group, 2012: 187).

The need to maintain and enhance biodiversity in food systems to ensure resilience against climate change is widely recognized, as is the need for change in industrialized agriculture to protect that biodiversity. The UN Food and Agriculture Organization states that:

> Production practices based on a continuing and increasing dependence on external inputs such as chemical fertilizers, pesticides, herbicides and water for crop production and artificial feeds, supplements and antibiotics for livestock and aquaculture production need to be altered. They are not sustainable, damage the environment, undermine the nutritional and health value of foods, lead to reduced function of essential ecosystem services and result in the loss of biodiversity (FAO, 2011: vii)

Likewise, the first ever International Assessment of Agricultural Knowledge, Science and Technology for Development, which was cosponsored by FAO, Global Environment Facility, UNDP, UNEP, UNESCO, WHO, and the World Bank, and was approved by 58 governments in 2008, asserts that:

> the loss of biodiversity and its associated agroecological functions (estimated to provide economic benefits of $1,542 billion per year) adversely affect productivity especially in environmentally sensitive lands in sub-Saharan Africa and Latin America. ... There have also been negative impacts of pesticide and fertilizer use on soil, air and water resources throughout the world. (IAASTD, 2009: 59)

The risk is clear. Industrialized agriculture is leading to a narrowing of the genetic foundation of our food system and poses a threat to the food security of future generations. The challenge is how to halt the decline and reduce the threat.

Agroecology

Agroecological food production systems are cited as one approach to addressing the loss of biodiversity and the consequent unsustainability of industrialized food production. Agroecological approaches recognize the interdependencies between our sources of food and the wider environment, and the overlapping needs to provide sustainable food systems and sustainable livelihoods. Agroecology promotes crop and livestock species diversity in order to exploit different ecosystem niches. This, in turn, provides greater system resilience to environmental change through the different genetic traits in breeds, for example, resistance to particular pathogens. It also supports future adaptive capacity by providing the foundation for further genetic variation, through breeding or variety selection, to meet as yet unknown challenges (Ensor, 2009). Resilience is further enhanced by reducing the need for external inputs, thus reducing costs and some of the risks associated with market volatility. Box 5.1 summarizes agroecological approaches that build resilience.

Box 5.1 Agroecological approaches that build resilience

Complex systems: In traditional agroecosystems the prevalence of complex and diversified cropping systems is essential to the stability of peasant farming systems, allowing crops to reach acceptable productivity levels in the midst of environmentally stressful conditions. Traditional agroecosystems are generally less vulnerable to catastrophic loss because they feature a wide range of crops and varieties in various spatial and temporal arrangements.

Use of local genetic diversity: In most cases, farmers maintain diversity as insurance against environmental change or to meet social and economic needs. The existence of genetic diversity has special significance for maintaining and enhancing productivity of small farming systems, as diversity also provides security against diseases, especially pathogens that may be enhanced by climate change. By mixing crop varieties, farmers can reduce the spread of disease-carrying spores and modify environmental conditions so that they are less favourable to the spread of certain pathogens, thus delaying the onset of diseases.

Soil organic matter enhancement: Throughout the world, small farmers use practices such as crop rotation, composting, green manures and cover crops, and agroforestry, which are all practices that increase biomass production and therefore build active organic matter. Soil management systems that maintain organic matter levels in the soil are essential to the sustained productivity of agricultural systems in areas frequently affected by droughts.

Multiple cropping or polyculture systems: Studies suggest that more diverse plant communities are more resistant to disturbance and more resilient to environmental perturbations. Intercropping, which breaks down the monoculture structure, can provide pest control benefits, weed control advantages, reduced wind erosion, and improved water infiltration.

Agroforestry systems and mulching: Many farmers grow crops in agroforestry-designed shade tree cover to protect crop plants against extremes in the microclimate and soil moisture fluctuation. Farmers influence the microclimate by retaining and planting trees, which reduce temperature, wind velocity, evaporation and direct exposure to sunlight, and intercept hail and rain. It is internationally recognized that agroforestry systems contribute simultaneously to buffering farmers against climate variability and changing climates, and to reducing atmospheric loads of greenhouse gases because of their high potential for sequestering carbon.

Source: Ensor, 2009: 12 (adapted by Ensor from Altieri and Koohafkan, 2008, and Altieri, 2002)

Traditional forms of agriculture tend to be agroecological in nature – localized, exploiting specific niches, and resulting in a wide variety of crops and species of livestock. These traditional varieties, many of which have been bred for particular traits, such as ripening time or winter hardiness, often represent the outcome of generations or sometimes millennia of farmer innovation in specific ecological environments. The case study from Sri Lanka of traditional rice varieties being used to meet the challenge of rising levels of salinity in coastal soils is but one example of the value of such traditional knowledge. There are many more. These include the return to traditional practices of intercropping coffee with indigenous trees in Peru to improve yields and restore rainforest cover (Henderson, 2014); the practice of traditional forms of aquaculture, where different local breeds of fish are kept together to exploit different ecological niches and enable optimum use of resources in a breeding pool; and the maintenance of indigenous breeds of livestock known for their ability to withstand particular diseases, such as the Red Masai sheep of East Africa, which is known for its resistance to intestinal worms (Ensor, 2009).

Indeed, as the FAO points out, 'local knowledge and culture can therefore be considered as integral parts of [agricultural biodiversity], because it is the human activity of agriculture that shapes and conserves this biodiversity' (FAO, 2004).

Restoring biodiversity and building resilience

Industrialized agriculture is a technology injustice in that it represents the use of a class of technology and technical knowledge that imperils global and local food systems and leaves future generations at risk of food insecurity. By promoting uniformity, it puts present-day food systems at risk from pandemic outbreaks of disease in large reservoirs of genetically uniform livestock or crops. By displacing traditional breeds and varieties, it reduces the genetic base of the food chain and narrows the gene pool for meeting future, as yet unknown, challenges. In addition, large-scale monocultures and the use of pesticides and herbicides impact on the wider environment and ecology, for example by the removal of vital pollinators or the destruction of fish and other species in our natural waterways.

The question arises, who will be the best steward of the genetic resources of our food systems through climate change and other challenges? The industrial food system whose breeding programmes have increased uniformity and vulnerability, or the tens of millions of peasant farmers around the world, like the rice farmer in Sri Lanka at the start of this chapter, who hold on to and understand the role that genetic diversity plays in securing their food security and livelihoods? And where should the collective effort to improve technology and technical knowledge go? Should public funding, subsidies, and food policies continue to support technological development in the industrialized food system, to aid its continued quest for productivity and uniformity of product? Or should it instead focus on the building of technical capabilities of the peasant farmers who currently maintain most of the genetic diversity in our food chain? Those technical capabilities relate to the conservation and use of diverse local species and agroecological and sustainable approaches to improve productivity that are vital to improving their food security and livelihoods while, at the same time, securing the genetic material needed to maintain global food supplies in the face of environmental and other challenges. The principle of Technology Justice would favour the latter, supporting the spread of technologies and technical knowledge that will benefit the largest number of people the most, not just today but in the future, too.

Energy security and climate change

> Senior executive, BHP Billiton (one of the largest global mining companies): *Well mate … we have good quality coal assets, close to market. And we also have gas resources, unlike most of our competitors. We're well diversified. So where is our problem exactly, d'ya reckon?*
>
> Jeremy Leggett, carbon campaigner and Chair of Carbon Tracker: *I look at him, attempting a poker face, processing this. His response is consistent with what we are finding with other companies. Their first line of defence is not to question the carbon arithmetic … (that to have a reasonable chance of keeping global warming below the intended two-degree ceiling, fully 80 per cent of existing fossil fuel reserves would have to stay in the ground unburned). It is to argue that their reserves are in the 20 per cent. I continue, 'We have noticed an interesting thing, in presentations like this. The first line of argument tends not to be a defence on behalf of your entire industry, but one specific to the company – that your particular set of assets can be burned safely. You can't all be right, can you?'* (Leggett, 2016)

Burning up the future

This exchange between the campaigner Jeremy Leggett and a senior executive of one of the world's largest coal mining companies encapsulates why our addiction to the use of fossil-fuelled technology is an injustice to future generations and how big the challenge to shake that addiction will be.

Fossil fuels currently account for 80 per cent of global energy consumption (IEA, 2015a). Figure 5.1 shows the breakdown of the sources of anthropogenic greenhouse gas emissions, generally recognized as the primary driver of climate change. In rough terms, emissions from the burning of fossil fuels account for two-thirds of total global emissions (with the remainder coming from land-use change and forestry, the management of agricultural soils and livestock, and methane from waste landfill sites).

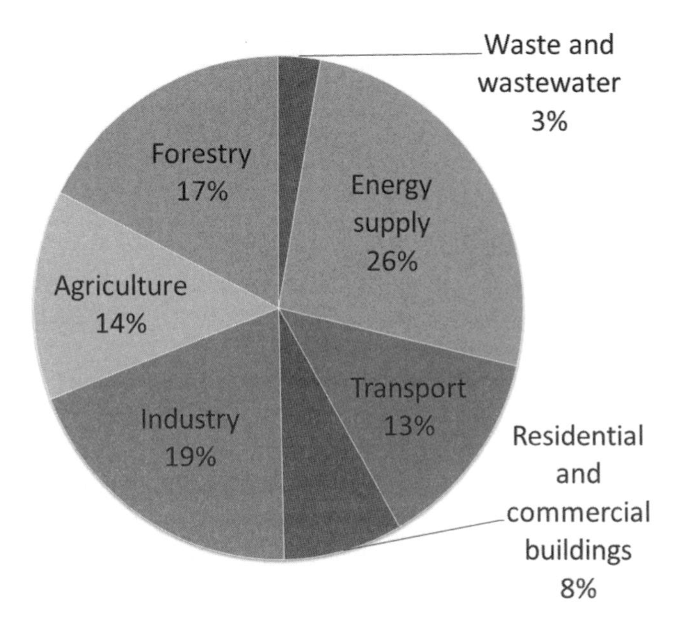

Figure 5.1 Proportion of global anthropogenic greenhouse gas emissions by source
Source: IPCC, 2007

Table 5.1 and Figure 5.2 portray how future risk changes as global mean temperatures rise. Table 5.1 shows the severity of potential impacts. The first column covers the level of risk to unique and threatened ecosystems while the second covers exposure to risks from extreme climate events, for example, increased frequency and/or intensity of storms, floods, and droughts. Figure 5.2 shows two of the IPCC's current emission change scenarios – a low and a high

Table 5.1 Severity of potential impacts of global temperature rise

Global mean temperature change (°C relative to 1986–2005)						
	4–5	Very high	High – Very high	Very high	High	High
	3–4	Very high	High	High – Very high	High	High
	2–3	Very high	High	High	Medium–High	High
	1–2	High	High	Moderate–High	Medium	Medium–High
	0–1	Moderate–High	Moderate	Moderate	Undetectable–Medium	Undetectable–Medium
Threats from climate change		*Unique and threatened systems*	*Extreme weather events*	*Distribution of impacts*	*Global aggregate impacts*	*Large-scale singular events*

Source: IPCC, 2014

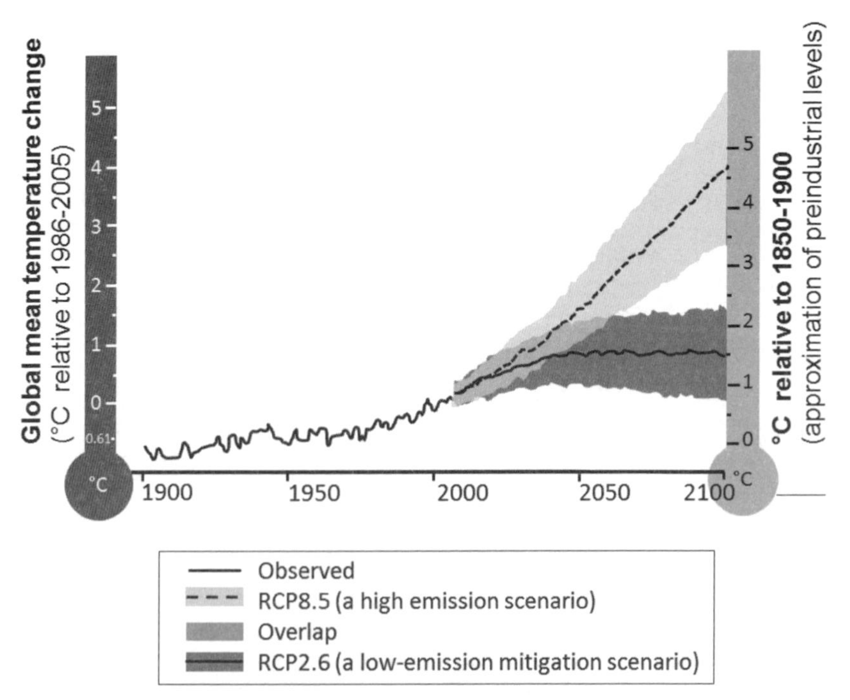

Figure 5.2 High and low-emission scenarios and impact on global temperature rise
Source: IPCC, 2014

one, for the period up to 2100. The shading indicates the range of uncertainty for each of the scenarios.

The IPCC concludes that:

> Increasing magnitudes of warming increase the likelihood of severe, pervasive, and irreversible impacts. Some risks of climate change are considerable at 1 or 2°C above pre-industrial levels. Global climate change risks are high to very high with global mean temperature increase of 4°C or more above pre-industrial levels ... and include severe and widespread impacts on unique and threatened systems, substantial species extinction ... [and] large risks to global and regional food security ... The precise levels of climate change sufficient to trigger tipping points (thresholds for abrupt and irreversible change) remain uncertain, but the risk associated with crossing multiple tipping points in the earth system or in interlinked human and natural systems increases with rising temperature. (IPCC, 2014: 14)

In human terms, as Jonah Busch of the Centre for Global Development says, 'Climate change is regressive – awful for the rich, but catastrophic for the poor' (Busch, 2014). Poor people live in locations more vulnerable to climate-induced disasters. They are dependent on farmlands that lack irrigation and are already more vulnerable to drought, or crammed into urban pockets that are not developed for middle-class housing or commercial purposes precisely because of susceptibility to damage from flooding or landslides, for instance. Their housing and other infrastructure is weaker and more likely to be destroyed in extreme weather events. Their access to insurance is limited, making post-disaster recovery more difficult. As the IPCC notes:

> Throughout the 21st century, climate-change impacts are projected to slow down economic growth, make poverty reduction more difficult, further erode food security, and prolong existing and create new poverty traps, the latter particularly in urban areas and emerging hotspots of hunger ... Climate-change impacts are expected to exacerbate poverty in most developing countries and create new poverty pockets in countries with increasing inequality, in both developed and developing countries. In urban and rural areas, wage-labor-dependent poor households that are net buyers of food are expected to be particularly affected due to food price increases, including in regions with high food insecurity and high inequality (particularly in Africa) ... (IPCC, 2014: 20)

Clearly, if the more extreme impacts of climate change are to be avoided, the world must decarbonize its energy systems rapidly to stabilize the amount of carbon dioxide in the atmosphere and keep global warming within the 2°C 'safe' limit. There are glimmers of hope that some progress is being made: despite rising energy use, for the first time in four decades, global carbon emissions associated with energy consumption remained stable in 2014 while the global economy grew (REN21, 2015). The renewable energy policy network REN21 attributed the stabilization to increased penetration of renewable energy and to improvements in energy efficiency. REN21 also noted that renewables accounted for nearly 60 per cent of the net additions to global power capacity during 2014, exceeding the combined coal and gas capacity added during the year.

There was also a lot of interest generated around news that renewable energy might be reaching 'grid parity' – the point at which it costs the same as energy from conventional gas or coal-fired power stations (oil is no longer a significant contributor to electricity generation in big energy-consuming economies such as the US and western Europe). In certain states in the US, electricity from onshore wind farms and larger solar arrays can undercut the price of fossil fuel-generated electricity, with unsubsidized wind costs falling from a minimum of $101 per megawatt hour (MWh) in 2009 to $37/MWh in 2014, while solar farm costs fell even more steeply from $323/MWh to $72/MWh (Murray, 2014). These costs compare favourably to power from new gas-fired power plants, which costs $61–87/MWh (Murray, 2014). Meanwhile, Deutsche Bank has predicted solar will reach grid parity in 80 per cent of world electricity markets by 2017 (Parkinson, 2015).

Yet, many challenges to decarbonization remain. The extent to which renewables can penetrate the electricity-generation market is limited by one major technical problem: storage. Electricity generation from wind and sun depends on the weather (and, in the case of solar, is limited to daylight hours). As electricity from renewables cannot yet be stored at scale, other conventional power sources are still needed to iron out intermittency fluctuations in renewable supplies and fine-tune the changes in balance between supply and demand during the day. Furthermore, electricity generation accounts for only 26 per cent of global greenhouse gas (GHG) emissions (see Figure 5.1) and the market penetration of renewable energy into power for other sectors such as transport, residential and commercial buildings, and industry has been much less marked.

It is not surprising that the IEA reports the long-term trend in GHG emissions has been relentlessly upwards since 1990, closely tracking global rises in gross domestic product. We are thus a long way from decarbonizing the global economy or delinking economic growth from growth in emissions. On a global level, both the energy intensity of GDP and the carbon intensity of primary energy have to be reduced by around 60 per cent by 2050 for us to have a hope of staying within the 2°C warming limit. But, at the moment, 'advances in those areas that were showing strong promise – such as electric vehicles and all but solar photovoltaics in renewable power technologies – are no longer on track to meet 2°C targets' (IEA, 2015b: 2).

Technology development and market economics are beginning to make large-scale solar and wind competitive with conventional power generation, as noted earlier. Advice from the IEA on the policy reforms that could hasten progress, ranging from the retirement of coal-fired power stations to the revival of the Emissions Trading System (IEA, 2014) is also to be welcomed. Internationally agreed targets can also help and the SE4All initiative has proposed three global targets to be achieved by 2030: universal access to energy, a doubling of the rate of increase in energy efficiency, and a doubling of the proportion of renewables in the global energy mix (SE4All, 2015). It is also encouraging to see that these three targets have been incorporated into the new Sustainable Development Goals (SDGs), although it is troubling that the renewables goal has been diluted to a non-specific commitment to 'substantially increase' rather than double the proportion of renewables in the global energy mix (UN, 2015: 2). And, of course, there are the continuing climate change talks, which showed significant signs of revival at the Paris COP21 meeting in 2015 with the final agreement committing to:

- ensure greenhouse gas emissions peak as soon as possible and, in the second half of this century, achieve a balance between sources and sinks of greenhouse gases;
- keep global temperature increase 'well below' 2°C and pursue efforts to limit it to 1.5°C;
- review progress every five years;
- provide $100 bn a year in climate finance for developing countries by 2020, with a commitment to further finance in the future (BBC, 2015).

Moreover, at the summit, and in parallel with the official negotiations, a thousand cities and more than 50 of the world's biggest companies committed to going 100 per cent renewable, some as early as 2030, while investors with funds worth more than $3 tn pledged to divest from fossil fuels and/or put shareholder pressure on traditional energy companies to decarbonize (Leggett, 2016).

However, despite the optimism emanating from the SDG and COP negotiations at the end of 2015, the world is still a long way from a safe transition to a post-carbon era. The SDGs are aspirational goals and there are no mandatory mechanisms to force governments to adjust national plans to fulfil the global commitments made. Likewise, the key pledges made at COP21 to reduce carbon emissions, the 'Intended Nationally Determined Contributions' (INDCs), are not legally binding but voluntary, based on what governments are prepared to deliver as opposed to what science demands. Moreover, the INDC pledges, when put together, would limit the global temperature rise by the end of the century to 2.7–3°C, well above the stated 1.5–2°C goal (Climate Action Tracker, 2015). Finally, the financial pledge of $100 bn a year in climate finance for developing countries by 2020 looks insignificant compared to estimates that the requirement is closer to $645–945 bn a year (see, for example, Montes, 2013; Chivers and Worth, 2015).[4]

It is clear that market-driven progress, global target setting, good policy prescriptions, and post-Kyoto climate negotiations have not yet set us on a path to keep global warming under 2°C. Indeed, as the exchange between Jeremy Leggett and the senior BHP Billiton executive illustrates, a major impediment to progress remains the sheer weight of vested interests in fossil fuels. These vested interests from the old oil, gas, and coal industries remain so large and so powerful that they can still persuade governments to maintain their licence to operate: a licence to drill for oil in the Arctic, to use fracking to exploit shale gas, to strip-mine large swathes of Alberta Province in Canada for oil sands, or to attempt to exploit frozen methane hydrates off the coast of Japan, despite the massive potential negative environmental risks associated with each of these ventures.[5] A licence to persist in doing this despite the knowledge that 80 per cent of existing proven oil reserves will have to remain in the ground unburned if we are to stay within the 2°C limit on global warming. And, most amazingly of all, a licence that costs taxpayers around the world more than $5 tn a year in post-tax subsidies (IMF, 2015), around 40 times the $120 bn equivalent global subsidy to renewables (Carrington, 2015).

The reason the continued use of fossil fuel is not merely an unfortunate necessity but is a technology injustice is that we, the present generation occupying this planet, have a choice. We do not have to continue to bow to the vested interests of the fossil fuel industry and we do not have to accept that the current rate of technological advancement on renewables and energy storage is good enough. It will require action of a financial nature, alongside global targets and policy prescriptions, to make a difference. Specifically, two things need to happen·

Firstly, the ridiculous global subsidy of $5 tn a year needs to be removed from fossil fuels. We cannot continue to subsidize an industry that we know is the major driver of climate change and which is resorting to ever more environmentally risky sources of supply for its key products. Removing the subsidy alone would, the IMF estimates, reduce global CO_2 emissions by 20 per cent (IMF, 2015: 26) and reallocating a portion of it to renewables could hasten improvements in energy efficiency, drive down the cost of renewables, and speed up the development of solutions to the energy storage problem.

Secondly, the campaign for disinvestment from the fossil fuel industry has to grow. As mentioned, an estimated 80 per cent of listed carbon reserves owned by companies will become 'stranded assets', having to stay in the ground unburned, if we are to have an 80 per cent chance of staying below a temperature rise of 2°C (Carbon Tracker, 2011: 2). This problem will get worse if current investment trends continue, as an estimated $674 bn of capital investment flows into exploration for new assets every year (Carbon Tracker, 2013: 13). In effect, a massive carbon bubble is being created which threatens both the global economy and the value of investments made by individuals, pension funds, and others in oil exploration companies. One way to defeat the vested interests is, simply, to educate the markets and make it increasingly clear just how risky it is to continue to invest in fossil fuels. Organizations such as the Carbon Tracker Initiative seek to provide the information needed for transparency in this area. The movement for institutions and governments to disinvest from fossil fuels is growing, further aided by groups such as Fossil Free, whose website lists the growing number of institutions (31 universities, 44 cities, 86 religious bodies, 34 foundations, and 28 other organizations in July 2015) that have already committed to disinvest from the industry. Recent high-profile successes by the campaign have included the government of Norway, which bowed to pressure in May 2015 and committed to disinvest its sovereign wealth fund of all investments in coal (New York Times, 2015).

Only a sustained and serious attack of this kind on the finances of the fossil fuel industry, combined with a massive increase in support to renewables, is likely to deliver transformation swift enough to avoid disastrous climate change and an immense technological injustice being perpetrated by current generations on future ones.

The misuse of antibiotics: turning the clock back for medicine

> *The time may come when penicillin can be bought by anyone in the shops. Then there is the danger that the ignorant man may easily under-dose himself and by exposing his microbes to non-lethal quantities of the drug make them resistant. Here is a hypothetical illustration. Mr X has a sore throat. He buys some penicillin and gives himself not enough to kill the streptococci but enough to educate them to resist penicillin. He then infects his wife. Mrs X gets pneumonia and is treated with penicillin. As the streptococci are now resistant to penicillin the treatment fails. Mrs X dies. Who is primarily responsible for Mrs X's death? Why, Mr X whose negligent use of penicillin changed the nature of the microbe. Moral: If you use penicillin, use enough.*

Extract from Sir Alexander Fleming's Nobel Prize speech on the discovery of penicillin

Anne Sheafe Miller was lucky. In March 1942, she was close to death at a hospital in New Haven, Connecticut, suffering from a streptococcal infection, an often fatal disease at that time. In hospital for nearly a month, delirious and with a temperature regularly reaching 41.5°C, there seemed little hope. Doctors had tried sulphonamide drugs, blood transfusions, and surgery, but nothing worked. As a last resort, her doctors obtained a small amount of penicillin (still a little-known experimental drug) and injected her with it. Within hours her temperature had dropped sharply and within a day she was no longer delirious and was soon eating again (Saxon, 1999). She was the first patient to be successfully treated with penicillin and went on to live to the age of 90.

Today, death from a minor scratch seems unimaginable but this was a common occurrence before antibiotics were discovered. Alarmingly, just a few years after Miller passed away in 1999, it looks like the antibiotic, a medical technology that has transformed health-care and life expectancy around the world, may already be coming to the end of its useful life. Sir Alexander Fleming's warning has come to pass. The careless use of antibiotics has indeed changed the nature of the microbe and, increasingly, bacteria are becoming resistant to other antibiotics and antimicrobials as well as penicillin. The race is on to prevent this from becoming another intergenerational technology injustice.

The scale of the problem

In 2014 the World Health Organization released a report on antimicrobial resistance across 115 countries, which cited among its evidence the rising resistance to the antibiotic carbapenem (a treatment of last resort for life-threatening infections caused by a common intestinal bacterium, *Klebsiella* pneumonia), to fluoroquinolones (one of the most widely used antibacterial medicines for the treatment of urinary tract infections caused by *E. coli*), and to cephalosporins (the treatment of last resort for gonorrhoea) (WHO, 2014a). The accompanying press release stated:

> this serious threat is no longer a prediction for the future; it is happening right now in every region of the world and has the potential to affect anyone, of any age, in any country. Antibiotic resistance – when bacteria change so antibiotics no longer work in people who need them to treat infections – is now a major threat to public health. (WHO, 2014b)

A review commissioned by the UK government confirms that antimicrobial-resistant bacteria already claim 50,000 lives in Europe and North America every year (Review on Antimicrobial Resistance, 2014). Estimates of the worldwide death toll are difficult to make given the lack of accurate statistics, but the review suggests it could be higher than 700,000 per year. As Figure 5.3 shows, if new antimicrobials are not found, the annual death toll could reach 10 million by 2050, dwarfing the morbidity rates of many of the major disease groups today.

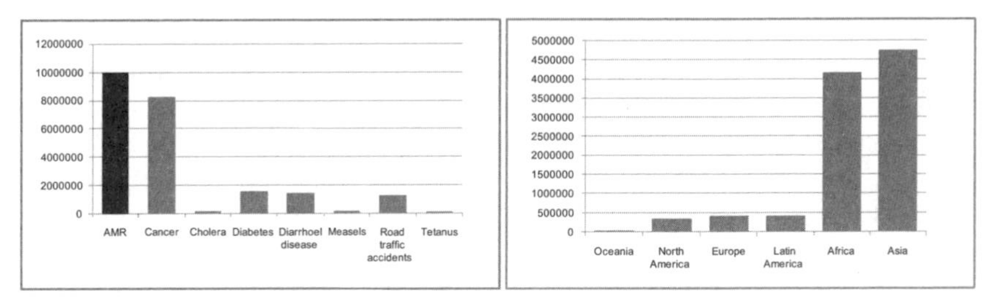

Figure 5.3 Predicted global annual death rates resulting from antimicrobial resistance compared to other diseases in 2015

Source: Adapted from Review on Antimicrobial Resistance, 2014:5

The causes of resistance

The mutation and evolution of bacteria and viruses is a natural process in response to changing environmental stressors. Penicillin and subsequently discovered antimicrobials have thus always had an inbuilt expiry date, as the bacteria they have been developed to fight will develop resistance as they evolve over time. But careless use of these valuable resources has dramatically shortened their useful lifespans.

The primary cause of this accelerated development of resistance has been overuse and misuse of antibiotics and other antimicrobials in two areas: human health and animal health. The widespread inappropriate prescription of antibiotics in the developed world has been well documented. A

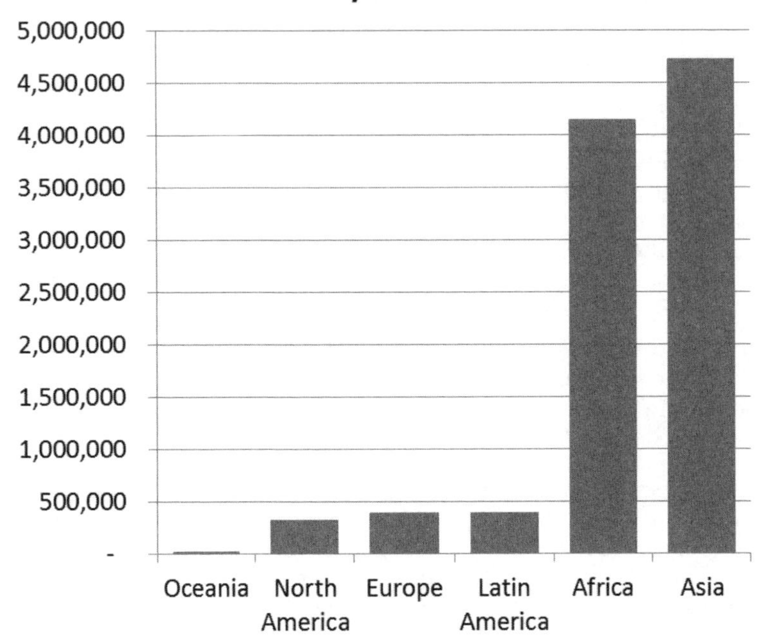

Figure 5.4 The geographical spread of morbidity in 2050
Source: Adapted from Review on Antimicrobial Resistance, 2014:13

year-long study in California in 1998–99 of more than 33,000 patients presenting to doctors with symptoms of the common cold or upper respiratory tract infections recorded that just under 36 per cent of them were prescribed antibiotics (Gilberg et al., 2003). That is despite the fact that these illnesses are caused by viruses, not bacteria, and so cannot be treated with antibiotics. The rate at which antibiotics are prescribed varies significantly across different countries even within Europe, indicating varying levels of laxity in medical practice. The OECD collects data on prescribing in different countries using a system of average defined daily doses to account for the volume of drugs. For the period 2000 to 2010, the defined daily dose of antibiotics for systemic use per 1,000 people in Greece was consistently three to three and a half times as much as in the Netherlands, as one example (Quality Watch, 2015). Increased use of surface antibacterial agents in many household products has also been cited as a contributor to the problem (Levy, 2002).

In the developing world, health systems are under huge stress and often unable to enforce regulation or offer the sort of routine drug-sensitivity testing and surveillance required to make sure antibiotics and other antimicrobial drugs are rationally used. Cheap, broad-spectrum drugs are often favoured, which fosters widespread resistance. Poor people often cannot afford to consult properly qualified diagnosticians and prescribers. Even when they are able to get proper advice they may lack the money to buy a full course of antibiotics, buying and consuming partial doses instead, something the lack of regulation in many developing countries facilitates. Substandard medicines also circulate in these markets, meaning that many antibiotics used do not contain the full dose required. This combination of overuse of partial doses of weak antibiotics provides an ideal breeding ground for rapid evolution of drug-resistant bacteria (Okeke, 2010).

Overuse of antibiotics is not limited to humans. They are also used excessively in animals and not only as veterinary drugs for the treatment of diseases, but also as prophylactics for disease prevention and, especially in the developed world, as growth promoters. Up to 80 per cent of the antibiotics used in the US are fed to animals, including 30 different types provided in low doses in livestock feed and water, 18 of which the US Food and Drug Administration allowed despite rating them as of 'high risk' for introducing antibiotic-resistant bacteria into the human food supply (*The Scientist,* 2014). The WHO notes that:

> The classes of antibiotics used in food-producing animals and in human drugs are mostly the same, thereby increasing the risk of emergence and spread of resistant bacteria, including those capable of causing infections in both animals and humans. Food-producing animals are reservoirs of pathogens with the potential to transfer resistance to humans. The magnitude of such transmission from animal reservoirs to humans remains unknown, and will probably vary for different bacterial species. The spread of resistance genes from animal bacteria to human pathogens is another potential danger which adds complexity. (WHO, 2014a: 59)

A further factor exacerbating the problem has been the lack of development of new antibacterial drugs in recent years. The 'golden era' of antibacterials was the 30-year period between 1930 and 1960 when the majority of the 22 distinct classes of antibacterial drugs were identified. Only five new classes have been discovered in the last 45 years, with the last being in the 1980s; since when there has been a 'discovery void' (WHO, 2014a: 1).

The nature of the injustice

It is easy to get absorbed in the detail of the multiple causalities for the growing problem of antimicrobial resistance and so lose sight of the nature and scale of the injustice this represents. It is important to remember that prior to the 1950s infections were often incurable and a common cause of death. Without antibiotics, minor accidents – perhaps just a scratch picked up while gardening – could prove fatal.[6] Without the prophylactic use of antimicrobials to prevent infection, any hospital operation, even one that today would be considered minor, carried with it the threat of death. Venereal diseases and other bacterial infections were largely untreatable.

It is generally accepted that, alongside improving access to safe water and sanitation and the introduction of vaccines, antibiotics have been the major technological advance in medicine that has delivered a phenomenal impact on human health and massive reductions in mortality rates over the last century. It is also well known, as noted earlier, that new antibiotics have proved very difficult to find since the 1960s. Given that knowledge, Technology Justice, or justice as fairness, demands that today's generation uses antibiotics in a way that recognizes their importance to human health and that extends their useful working life for as long as possible, so the health benefits available from them now can be enjoyed by subsequent generations, too. The use of antibiotics today does anything but that. In addition to widespread inappropriate prescription of antibiotics in the developed world, the failure to provide affordable access to WHO's list of essential medicines (which includes antibiotics) has led to widespread under-dosing in the developing world, while, in agriculture, the use of antibiotics as growth promoters confirms that commercial profit from industrialized meat production is prioritized over human health.

Although it is inevitable that the efficacy of antibiotics will decline over time, as the bacteria they are designed to fight evolve, the speed at which that is happening is due to the lack of governance and oversight of the use of the technology. If a better way to regulate the use of antimicrobials cannot be found and research into their replacement cannot be escalated, then this generation will turn the clock back on much of the medical progress made in the last century and condemn future generations to a life less healthy and 10 million more deaths a year due to antimicrobial resistance by 2050.

The path ahead

This is a massively complex problem which requires work on many fronts. Broadly speaking, the research suggests three, not necessarily mutually exclusive, paths of action.

The first is to redouble efforts to develop new classes of antibacterial drugs that can replace those which are no longer effective. One of the major concerns here is the lack of volume of new compounds under development at the moment. The UK government's review on microbial resistance sees this as a problem of market correction. Despite growing resistance to the carbapenem class of drugs – medicine's last effective line of defence against infections of the bloodstream and pneumonia – as of May 2015 there were only three compounds under development that have the potential to replace the carbapenems. The pharmaceutical industry is failing to invest enough in development. The explanation given for this market failure is that:

> the commercial return for any given new antibiotic is uncertain until resistance has emerged against a previous generation of drugs. In other medical fields, a new drug is meant to significantly improve on previous ones and so will become the standard first choice for patients quickly

> once it comes to market. That is often not true for a new antibiotic: except for patients with infections that are resistant to previous generations of drugs, a new antibiotic is most probably no better than any existing and cheap generic product on the market. By the time that new antibiotic becomes the standard first line of care, it might be near or beyond the end of its patent life. This means that the company which developed it will struggle to generate sufficient revenues to recoup its development costs. (Review on Antimicrobial Resistance, 2015: 2)

The solution, according to the review, is a global innovation fund of around $2 bn over five years to remove the commercial risk and kick-start a higher volume of research.

The second path of action is, at least as a bridging activity, to tweak older drugs that have fallen out of use. One example has been work on an antibiotic called spectinomycin that was used in the 1960s to treat gonorrhoea. It had to be used in large doses and was eventually replaced by newer, more potent drugs. Using molecular chemistry techniques that were not available in the 1960s, scientists have managed to alter the structure of the compound to the extent that it was able to fight off tuberculosis in vitro and in mice (*The Scientist*, 2014).

The third is a more radical solution: to rethink entirely the use of antimicrobials in the first place. This would include fundamentally questioning the use of antibiotics in healthy animals as prophylactics and growth promoters. It would mean rethinking the extent to which we tackle the problem of infection from the perspective of a war against 'bad bacteria' or a pact to help support 'good bacteria'. The latter would involve reversing the resistance problem by reviving strains of bacteria that are susceptible to existing drugs but which can compete with and replace the bacteria that are now resistant to antibiotics. In effect this is the approach adopted by probiotics. 'Preempt', for example, is a commercially available product that prevents salmonella in chickens by introducing other bacteria strains from adult hens that can compete with and exclude the salmonella bacteria (Levy, 2002: 28).

It is most likely that a combination of the three approaches will be required, at least in the short term. To stand a chance of success they have to be accompanied by a massive public re-education effort to raise awareness and appreciation of the dangers of antibacterials and the severe consequences of the technological injustice that arises from their misuse. But ultimately a race of drug development against bacteria cannot be a long-term solution, as attested by the current classes of antibiotics being already close to obsolescence just a few years after the first person to be cured by them has passed away.

Stuart Levy of Tufts University of Medicine puts it well:

> We began the antibiotic era with a full-fledged attack on bacteria. It was a battle misconceived and one in which we cannot be the winner. We cannot destroy the microbial world in which we have evolved. The best solution now is to take a broader view of the microbial world. While focusing on the pathogens, our efforts should act in ways that impact

> fewer commensal flora (the microbes and bacteria that normally inhabit our bodies). We need to forget 'overcome and conquer' and substitute 'peace' when regarding the microbial world. The commensal organisms are, in fact, our allies in reversing the resistance problem. As they rebuild their constituencies, they will control the levels of resistance by out-competing resistant strains. Then, when bacteria or other microbes cause infections, they will be drug susceptible, and we will have the ... [means] to treat them effectively and successfully. (Levy, 2002: 29)

Technology Justice and use

The previous sections have considered three examples of intergenerational technology injustices. They depict a clear and pressing need to think beyond current forms of regulation and governance to deal with the wider long-term impacts of the introduction of new technologies on the environment and on future generations. All three are technologies that have radically changed humankind's prospects. Fossil-fuelled technologies made the industrial revolution possible and transformed material living conditions for billions of people. The shift of human resources from agriculture to industry has been supported, at least in the developed world, by industrialized food production, enabling an era of cheap food delivered by a fraction of the original rural labour force, which, in turn, has allowed (or forced, depending on your viewpoint) a migration of populations from rural to urban environments where services such as energy, water, health, education, and transport can be provided more economically. Completing the triptych, the discovery of antibiotics has transformed the management of infections, extended lifespans, and dealt with the challenge of disease transmission when large populations are brought together in crowded urban environments.

But these great leaps forward seem to have been built on foundations of sand. The fossil fuel era could be argued to have lasted from 1850 to the current day – just over one and a half centuries. Antibiotics have been in use as a treatment since the mid-1940s – maybe 70 years. The Green Revolution that ushered in the radical intensification of agriculture started in the 1960s – just half a century ago. And yet, already, all these technologies have created unprecedented levels of environmental and social problems, threatening an unjust and toxic legacy for future generations to struggle with. In a very short space of time, the technologies that have transformed the living standards and lifespans of at least two-thirds of people on the planet already appear obsolete, the unforeseen and unintended consequences of their use bringing humankind to the cusp of disaster.

Carrying on in this manner is not an option. Either we find a new way to govern the development and use of technology – a way which will allow us to make those compromises that Rawls's theory of justice requires, a way which will allow us to live within Rockström's planetary boundaries – or the planet's own 'antibacterial systems' of climate change, famine, and disease will in turn eject us as the 'bad bacteria' in the system.

Notes

1. A suction handpump has its piston in the pump body above ground. It works by creating a vacuum that draws water up a pipe from below ground. Atmospheric pressure (and the efficiency of the seal in the pump) sets the maximum depth that a suction pump can lift water – generally 7 m or less. If water needs to be lifted more than 7 m (when the water table is more than 7 m below ground), a different type of pump, which places the piston below ground and below the water table, needs to be used.
2. Author's recollection from the work of the NGO WaterAid in Bangladesh.
3. Taken from personal field trip notes made by the author during an interview with the farmer.
4. Based on the World Bank 2010 estimate of mitigation costs of $265–565 bn a year and the UNFCCC 2009 estimate of adaptation costs at a further $380 bn a year (Montes, 2013).
5. According to Greenpeace, 'the Arctic's extreme weather and freezing temperatures, its remote location and the presence of moving sea ice severely increase the risks of oil drilling, complicate logistics and present unparalleled difficulties for any clean-up operation. Its fragile ecosystem is particularly vulnerable to an oil spill and the consequences of an accident would have a profound effect on the environment and local fisheries' (Greenpeace, 2014). The British Geological Survey notes the 'potential detrimental environmental impacts from fracking for gas carbon dioxide (CO_2) and methane (CH_4) emissions, particularly the potential for increased fugitive CH_4 emissions during drilling compared with drilling for conventional gas, the volumes of water and the chemicals used in fracking and their subsequent disposal, the possible risk of contaminating groundwater, competing land-use requirements in densely populated areas, the physical effects of fracking in the form of increased seismic activity' (BGS, 2013). The environmental impact of destroying hundreds of square kilometres of boreal forest in Alberta to access oil sands, the potential risks from oil spills from the associated 1673-mile Keystone XL pipeline, which traverses highly sensitive terrain, and the higher GHG emissions associated with oil sands compared to conventional oil is well covered in Naomi Klein's book *This Changes Everything* (Klein, 2014). Environmentalists fear that drilling for frozen methane in the difficult deep-water conditions off the coast of Japan could lead to a leak of the greenhouse gas, which is 21 times as damaging as CO_2 (Fitzpatrick, 2010).
6. See the example of Albert Alexander, the first person to respond positively to penicillin after picking up an infection from scratching himself on a rosebush, but who died because insufficient supplies of the antibiotic were available to complete treatment (Saxon, 1999).

References

ActionAid (2010) *Meals per Gallon*, London: ActionAid.
ActionAid (2012) *Biofueling Hunger*, Washington, DC: ActionAid.

Altieri, M.A. (2002) 'Agroecology: the science of natural resource management for poor farmers in marginal environments', *Agriculture, Ecosystems and Environment*, 93(1–3): 1–24 <http://dx.doi.org/10.1016/S0167-8809(02)00085-3>.

Altieri, M.A. and Koohafkan, P. (2008) *Enduring Farms: Climate Change, Smallholders and Traditional Farming Communities*, Penang: Third World Network.

BBC (2015) 'COP21 Climate Change Summit reaches deal in Paris', 13 December, <http://www.bbc.co.uk/news/science-environment-35084374> [accessed 20 January 2016].

Berger, R. (2009) 'Participatory rice variety selection in Sri Lanka', *Participatory Learning and Action*, 60: 88–98.

BGS (2013) 'Potential environmental considerations associated with shale gas', <http://www.bgs.ac.uk/research/energy/shaleGas/environmentalImpacts.html> [accessed 15 July 2015].

Busch, J. (2014) 'Climate change is regressive', 1 April, *Center for Global Development*, <http://international.cgdev.org/blog/climate-change-regressive> [accessed 25 January 2015].

Carbon Tracker (2011) *Unburnable Carbon – Are the World's Financial Markets Carrying a Carbon Bubble?*, London: The Carbon Tracker Initiative.

Carbon Tracker (2013) *Unburnable Carbon 2013: Wasted Capital and Stranded Assets*, London: The Carbon Tracker Initiative.

Carrington, D. (2015) 'Fossil fuels subsidised by $10m a minute, says IMF', 18 May, *The Guardian*, <http://www.theguardian.com/environment/2015/may/18/fossil-fuel-companies-getting-10m-a-minute-in-subsidies-says-imf> [accessed 16 July 2015].

Chivers, D. and Worth, J. (2015) 'Paris Deal: Epic fail on a planetary scale', 12 December, *New Internationalist Magazine* <http://newint.org/features/web-exclusive/2015/12/12/cop21-paris-deal-epi-fail-on-planetary-scale/> [accessed 20 January 2016].

Climate Action Tracker (2015) 'Effect of current pledges and policies on global temperature', 7 December, *Climate Action Tracker* <http://climateactiontracker.org/global.html> [accessed 20 January 2016].

Ensor, J. (2009) *Biodiverse Agriculture for a Changing Climate*, Rugby: Practical Action Publishing.

ETC Group (2012) 'The ETC report: the poor can feed themselves', in P. Pojman, *Food Ethics*, Boston, MA: Wadsworth.

EBTP (2012) 'Food vs fuel', *European BioFuels Technology Platform* <http://www.biofuelstp.eu/food-vs-fuel.html> [accessed 20 November 2015].

FAO (2004) 'What is agrobiodiversity?', *Food and Agriculture Organization of the United Nations* <http://www.fao.org/docrep/007/y5609e/y5609e01.htm> [accessed 15 July 2015].

FAO (2011) *Biodiversity for Food and Agriculture – Contributing to Food Security and Sustainability in a Changing World*, Rome: FAO.

Fitzpatrick, M. (2010) 'Japan to drill for controversial "fire ice"', 27 September, *The Guardian*, <http://www.theguardian.com/business/2010/sep/27/energy-industry-energy> [accessed 16 July 2015].

Fossil Free (2015) <http://gofossilfree.org/> [accessed 15 July 2015].

Gilberg, K., Laouri, M., Wade, S., and Isonaka, S. (2003) 'Analysis of medication use patterns: apparent overuse of antibiotics and underuse of prescription drugs for asthma, depression, and CHF', *Journal of Managed Care & Specialty Pharmacy*, 9(3), 232–37 <http://dx.doi.org/10.18553/jmcp.2003.9.3.232>.

Greenpeace (2014) 'The dangers of Arctic oil', *Greenpeace International*, <http://www.greenpeace.org/international/en/campaigns/climate-change/arctic-impacts/The-dangers-of-Arctic-oil/> [accessed 15 July 2015].

Henderson, C. (2014) 'Agro-ecology: a powerful tool for adaptation to climate change and the prosperity of smallholder farmers', 8 December, *Practical Action*, <http://practicalaction.org/blog/news/agro-ecology-a-powerful-tool-for-adaptation-to-climate-change-and-the-prosperity-of-smallholder-farmers/> [accessed 14 July 2015].

IAASTD (2009) *Agriculture at a Crossroads: IAASTD Synthesis Report*, Washington, DC: International Assessment of Agricultural Knowledge, Science and Technology for Development.

IEA (2014) *Energy, Climate Change and Environment, 2014 Insights: Executive Summary*, Paris: International Energy Agency.

IEA (2015a) *Climate Change*, International Energy Agency, <http://www.iea.org/topics/climatechange/> [accessed 15 July 2015].

IEA (2015b) *Energy Technology Perspectives in 2015: Executive Summary*, Paris: International Energy Agency.

IMF (2015) *How Large are Global Energy Subsidies?*, Washington, DC: International Monetary Fund.

IPCC (2007) 'Climate change 2007: synthesis report', Intergovernmental Panel on Climate Change, <http://www.ipcc.ch/publications_and_data/ar4/syr/en/figure-spm-3.html> [accessed 15 July 2015].

IPCC (2014) *Climate Change 2014: Impacts, Adaptation and Vulnerability. Summary for Policy Makers*, Cambridge: Cambridge University Press.

Klein, N. (2014) *This Changes Everything: Capitalism vs the Climate*, New York: Simon and Schuster.

Leggett, J. (2016) *The Winning of the Carbon War: Power and Politics on the Front Lines of Climate and Clean Energy*, London: Leggett.

Levy, S.B. (2002) 'Factors impacting on the problem of antibiotic resistance', *Journal of Antimicrobial Chemotherapy*, 49: 25–30 <http://dx.doi.org/10.1093/jac/49.1.25>.

Montes, M. (2013) *Climate Change Financing Requirements of Developing Countries – Climate Policy Brief 11*, Geneva: South Centre.

Murray, J. (2014) 'Lazard: US renewables reach cost-parity tipping point', 18 September, *Business Green*, <http://www.businessgreen.com/bg/news/2370937/lazard-us-renewables-reach-cost-parity-tipping-point> [accessed 16 July 2015].

New York Times (2015) 'Norway divests from coal', 9 June, *New York Times*, <http://www.nytimes.com/2015/06/10/opinion/norway-divests-from-coal.html?_r=0> [accessed 16 July 2015].

Okeke, I. (2010) 'Poverty and root causes of resistance in developing countries', in A. de Sosa (ed.), *Antimicrobial Resistance in Developing Countries*, pp. 27–36, New York: Springer.

Parkinson, G. (2015) 'Deutsche Bank predicts solar grid parity in 80% of global market by 2017', 14 January, *Clean Technica*, <http://cleantechnica.com/2015/01/14/deutsche-bank-predicts-solar-grid-parity-80-global-market-2017/> [accessed 16 July 2015].

Quality Watch (2015) 'Antibiotic prescribing', *Quality Watch*, <http://www.qualitywatch.org.uk/indicator/antibiotic-prescribing#vis-ref_585> [accessed 16 July 2015].

REN21 (2015) *Renewables 2015 Global Status Report*, Paris:REN21/ United Nations Environment Programme.

Review on Antimicrobial Resistance (2014) 'Antimicrobial resistance: tackling a crisis for the health and wealth of nations', London: UK Government.

Review on Antimicrobial Resistance (2015) 'Securing new drugs for future generations: the pipeline of antibiotics', London: UK Government.

Saxon, W. (1999) 'Anne Miller, 90, first patient who was saved by penicillin', 9 June, *New York Times*, <http://www.nytimes.com/1999/06/09/us/anne-miller-90-first-patient-who-was-saved-by-penicillin.html> [accessed 17 July 2015].

The Scientist (2014) 'Overcoming resistance', 1 April, *The Scientist*, <http://www.the-scientist.com/?articles.view/articleNo/39512/title/Overcoming-Resistance/> [accessed 17 July 2015].

UN (2015) 'Draft outcome document of the United Nations summit for the adoption of the post-2015 development agenda', *Sustainable Development Knowledge Platform*, <http://www.un.org/ga/search/view_doc.asp?symbol=A/69/L.85&Lang=E> [accessed 26 March 2016].

SE4All (2015) *United Nations Sustainable Energy for All Global Tracking Framework*, New York: United Nations Sustainable Energy for All Initiative.

Wharton University (2013) 'The transportation nexus: ethanol is a "food vs. fuel" issue', 10 July, *Knowledge@Wharton*, <http://knowledge.wharton.upenn.edu/article/the-transportation-nexus-ethanol-is-a-food-vs-fuel-issue/> [accessed 20 November 2015].

WHO (2014a) *Antimicrobial Resistance – Global Report on Surveillance*, Geneva: World Health Organization.

WHO (2014b) 'WHO's first global report on antibiotic resistance reveals serious, worldwide threat to public health', 30 April, *World Health Organization*, <http://www.who.int/mediacentre/news/releases/2014/amr-report/en/> [accessed 17 July 2015].

CHAPTER 6

The governance of technology access and technology use: time for a rethink

At this point it is appropriate to draw together the threads of the arguments made in the first part of this book, which has focused on injustices in how existing technology is accessed and used, and on the relationship between technology and the dual challenges of ending poverty and achieving environmental sustainability.

Injustices in access to and use of technology

The idea of technological determinism is flawed. Neither the development path of technology nor its impact on society is inevitable. Social, cultural, and economic factors play an important role in influencing technological innovation processes and the extent to which technologies are disseminated among, assimilated, and used by different groups within societies. As shown in Chapter 3, the invention of the light bulb has not inevitably led to universal access to electricity, while a technical understanding of the aetiology of diarrhoeal disease and cholera has not inevitably led to universal access to safe water and sanitation, despite more than a century passing since the introduction of each technology.

The fact that more than 1 billion people today are still living in the dark, without access to electricity, that 700 million lack safe water, that 2.5 billion are denied safe sanitation, and that one in three people around the world cannot access WHO's list of essential medicines is neither 'unfortunate' nor 'inevitable'. It is the direct result of choices. Choices around how much priority is given to providing services to those who have none, over improvements to services for those who already have them; choices around the importance attached to achieving a universal basic social foundation; and choices about how such a foundation is financed and by whom, and with what technology it will be delivered. The failure to achieve that social foundation, when the technologies necessary are already available, is an injustice to the poorest in society today on a massive scale: a technology injustice.

The invention and dissemination of new technologies can be an unpredictable and somewhat haphazard process. Technologies are not, ultimately, always used in the way or for the purpose that was envisaged when they were developed. Likewise, the wider social, economic, and environmental impacts that technologies may ultimately bring generally cannot be fully anticipated prior to their adoption and widespread use. But, as the examples

http://dx.doi.org/10.3362/9781780449043.007

in Chapter 5 show, whether it is the use of corn ethanol to fuel cars in the US affecting the price of *tortillas* in Mexico, or the adoption of industrialized food production diminishing biodiversity in the food chain, our record of reflexivity – of changing course to avoid or minimize the negative impacts of technology use when problems are identified – is poor. The failure to anticipate at the outset of the Green Revolution the full impact of pollution that would arise from the use of fertilizers, pesticides, and herbicides, or the extent by which the search for uniformity and productivity through breeding would massively reduce biodiversity in food crops and livestock is understandable. And maybe humankind was just being too optimistic (or technologically deterministic) when it ignored Alexander Fleming's warning about how easily bacterial resistance to antimicrobials could grow and assumed we would continue to discover new compounds as older antibiotics became less effective. But it is simply a gross technological injustice to present and future generations that, with our knowledge of planetary boundaries, humankind should persist with the use of technologies it knows to be pushing the world past the carrying capacity of its ecological systems.

A single unified problem

Historically, poverty in the developing world and global environmental sustainability have been addressed as separate, if parallel, issues. This has been true in institutional terms (for example, in the civil-society sector where organizations are generally seen as members of the environment or the development lobby) and in process terms (for example, the Millennium Development Goals as opposed to the Convention on Biological Diversity). This division is not tenable. Population growth, globalization, and technology have entwined developed and developing nations in a net of relationships and feedback loops that cannot be disentangled. A mobile phone on sale in the UK may contain solder made from the tin mines that have caused environmental devastation on Indonesia's Bangka island (FOE, 2012) or the metal tantalum, mined under appalling labour conditions in the Democratic Republic of Congo (Fairphone, 2015). Once used and discarded it may end up as e-waste on a tip on the outskirts of Accra in Ghana, not only polluting the environment but also harming the health of poor people trying to work the dump to recycle old electronic devices (Goutier, 2014). A decision to promote biodiesel as a fuel for environmental reasons in Europe ends up displacing smallholder farmers in Africa as land is grabbed to feed the new demand. The impacts of climate change know no boundaries and although the developed world is responsible for by far the greatest portion of historical greenhouse gas emissions, how poverty is tackled in the developing world – whether through dirty or green growth – will be at least as important as how the developed nations reduce their emissions in determining whether the world stays within the 2°C boundary necessary to keep environmental impacts manageable.

Raworth's model of that safe space for humanity, between achieving a minimum social foundation for everyone and managing human activity to stay within safe planetary boundaries, recognizes this. The Sustainable Development Goals also recognize this. But this recognition is not reflected in the way technology dissemination and use is governed. A Rawlsian approach to justice would demand fairness in terms of access to technologies, but also compromise to ensure that access doesn't come at the cost of others, now or in the future, not having the same.

References

Fairphone (2015) 'Mining: providing alternatives for miners in conflict-affected regions', *Fairphone*, <https://www.fairphone.com/roadmap/mining/> [accessed 3 December 2015].

FOE (2012) 'Mining for smartphones: the true cost of tin', London: Friends of the Earth.

Goutier, N. (2014) 'E-waste in Ghana: where death is the price of living another day', 7 August, *The Ecologist*, <http://www.theecologist.org/News/news_analysis/2503820/ewaste_in_ghana_where_death_is_the_price_of_living_another_day.html> [accessed 3 December 2015].

PART II: RETOOL

Driving innovation to develop the right technologies

CHAPTER 7

The link between technological innovation and economic development

Justice as a fair space for innovation

From a Technology Justice perspective, it is vital to understand how innovation happens and thus what policy options exist to drive innovation towards addressing poverty and environmental sustainability. The notion of steering or helping accelerate the innovation process is particularly important given the short window of opportunity humankind has to avoid catastrophic climate change, the need to work simultaneously on environmental and poverty reduction goals, and the level of technological change that is required to achieve both.

Chapter 2 referred to the American philosopher John Rawls's idea of justice as fairness as a way of considering injustices in access to and use of technology. The German sociologist Rafael Ziegler shows it is possible to use Rawls's notion of fairness to look at technology innovation, too. He notes that:

> Innovation discourse tends to centre on winners, focusing on those ideas that reproduce, survive and 'succeed' ... Accordingly, from a perspective of justice, special attention is needed to include all innovations of relevance for justice in this selection process and to ensure that the process works in such a way that it promotes justice or at least does not increase injustice. Thus, there is an epistemic–ethical challenge to conceptualize innovation in such a way that ideas are not arbitrarily excluded from the outset or blocked in the process. Practically, we can formulate this as the challenge of creating a fair space for innovation. (Ziegler, 2015)

Within this concept, Ziegler sees two potential ways in which innovation itself could be just:

- It could contribute to the long-term stability of society by finding creative responses to societal challenges, such as climate change.
- It could focus on innovation that is specifically aimed at improving conditions for the least advantaged in the present.

Ziegler draws on innovation systems thinking and the recent writing of academics such as Mariana Mazzucato to note that technological innovation is not simply the result of private investments into entrepreneurial responses to markets, but is also the outcome of systemic interactions between a wide range of public and private institutions. These institutions create the general environment in which innovation happens and finance the basic research

http://dx.doi.org/10.3362/9781780449043.008

that companies then draw on. This, Ziegler argues, means that when trying to understand what a fair space for innovation might look like, we not only have to consider whether the impact of innovation is shared justly, but also whether its gains are justly distributed according to the public and private investments that were necessary to allow it to happen. To use the analogy of recent problems in the banking and finance sector, we need to ensure that in supporting technology innovation we avoid distributing a return on investment that results in the privatization of gain but the socialization of risk.

The following chapters of this book explore further the idea of a fair space for innovation and its relevance to Technology Justice. This chapter starts by looking at how economic theories, from neoclassical growth models to innovation systems approaches, have dealt with innovation, before looking at the implications of the latter for technology innovation in the developed world and technology transfer and 'technology catch-up' in less developed economies. Chapter 8 then looks at the twin governance challenges implicit in Technology Justice: managing the risks associated with the undesirable impacts of new technologies, and ensuring innovation efforts address society's key needs. Chapter 9 looks at the role intellectual property rights play in stimulating innovation, while Chapter 10 explores whether the state plays a far larger entrepreneurial role than traditionally thought in commercial innovation processes. Part II closes with Chapter 11, which poses questions about who innovation is for and who should be involved in it, and considers what innovation in the absence of commercial drivers might look like.

Understanding how innovation happens

Before discussing whether existing technology innovation processes are 'just' and whether there is indeed a 'fair space' for technology development, it is important to understand the evolution of thinking around how innovation happens and what drives it. It is this thinking that has influenced national government science and technology innovation policies in the developed world, and shaped ideas and practices around technology transfer and catch-up in developing countries for the last 50 years or so.

Technological innovation in neoclassical economic growth models

Technological innovation has featured in economic models since the 1950s as a factor influencing growth. That, in turn, has driven interest in understanding how innovation happens and what can be done to encourage more of it.

The economist Robert Solow is generally credited with introducing technology into what is known in neoclassical economics as the 'production function'. The production function is generally used to specify the maximum output obtainable from a given set of inputs. It allows economists to ignore managerial and technological challenges associated with realizing this theoretical maximum, and to focus instead on the problem of the efficiency of allocating resources: the economic choice concerning which combination of inputs can produce the maximum level of outputs. In the early 1950s, the conventional view was that production, and thus economic growth, was determined by the quantities of labour and capital inputs applied alone. Production curves could be drawn to show the relationship between inputs and outputs (see Figure 7.1). Analysis focused on how capital and labour could be substituted for each other to maximize growth.

Solow won a Nobel Prize in 1987 for his 'contribution to the theory of economic growth' and, more specifically, for his work in 1956–7 to challenge this view of the production function. As he noted in his acceptance speech, Solow was able to demonstrate that labour and capital inputs could actually account for only about 10 per cent of economic growth in advanced economies, such as that of the US (Solow, 1987). He proposed that the missing 90 per cent was accounted for by technology innovation, which allowed, for example, particular industrial processes to be carried out in a new way, using fewer resources. Technological innovation could change the production curve, lifting it so that more outputs were produced for the same input levels of other resources (Figure 7.2).[1]

Solow's model was seen to have problems, however, in that it treated the technological innovation part of the economic 'production function' as a bit of a 'black box', an external variable that just happened to companies and economies as opposed to something internal to the process of production. This was one of the reasons for the revival in interest of the works of the economist Joseph Schumpeter in the 1980s, in particular his view of the dynamic, turbulent, and uncertain processes involved in technological innovation outlined in his book *Theory of Economic Development* (Schumpeter, 2008 [1911]). Schumpeter described development as a historical process of structural change, driven largely by innovation. In his view, the invention of a new technology or process was much less important in terms of economic impact than its commercialization (Śledzik, 2013). As Solow himself reflected in 2007, Schumpeter's view was that:

> Innovation is not the same thing as invention. Anyone can invent a new product or a new technique of production. The entrepreneur is the one who first sees its economic viability, bucks the odds, fights or worms his way into the market, and eventually wins or loses. Each win means profit for the entrepreneur and his backers, and it also means a jog upward for the whole economy. In the course of this process, which cannot possibly run smoothly, many businesses, individuals, and institutions, themselves founded on earlier successful innovations, will be undermined and swept away. Schumpeter called this birth-and-death process 'creative destruction,' and realized before anyone else that it was the main source of economic growth. (Solow, 2007)

The result of Schumpeter's influence was that economists shifted to a more complex analysis of how innovation actually happened. Rather than treating

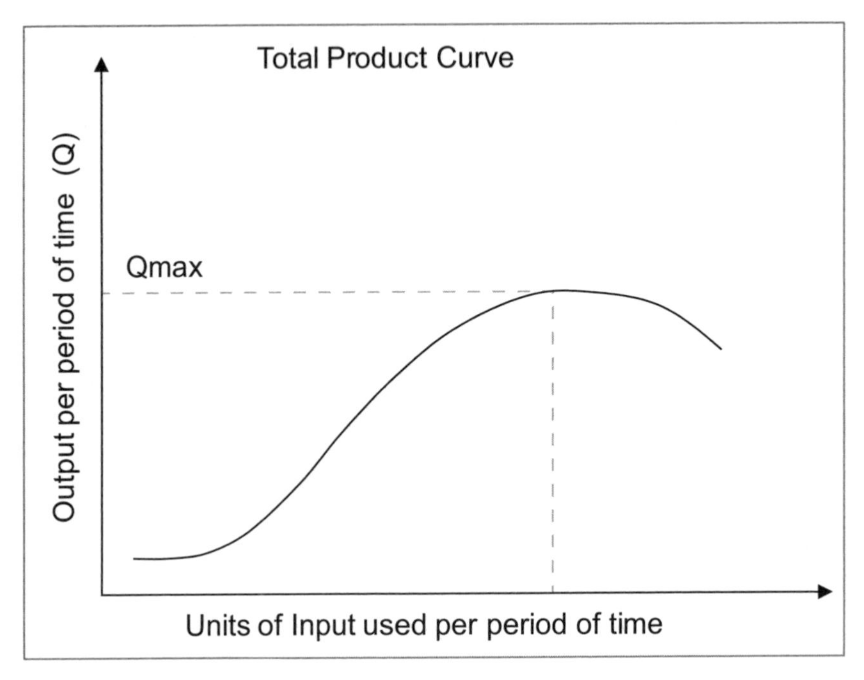

Figure 7.1 The production curve

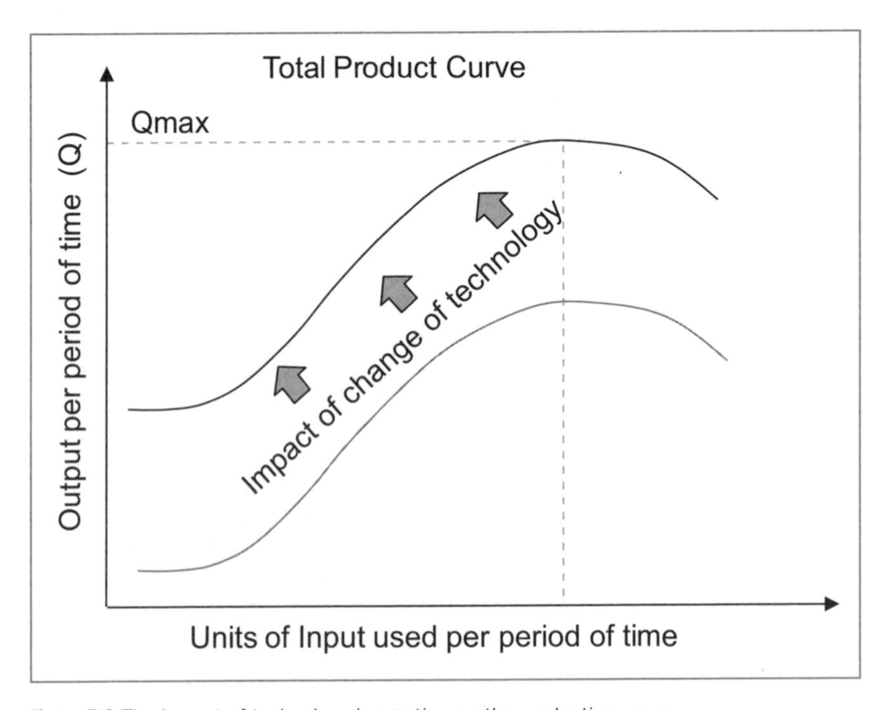

Figure 7.2 The impact of technology innovation on the production curve

technological innovation as a random occurrence that might happen anywhere in an economy, this body of thought highlighted the role of the entrepreneur in risk-taking (given the uncertainty involved in innovation) and in providing the leadership needed to move quickly to reap an economic reward before someone else and to overcome the inertia to new ideas inherent in society (Fagerberg, 2003). This approach was an attempt to account for the fact that technological innovation seemed to occur at different rates in different locations.

One approach based on Schumpeter's work has been to classify innovation according to how radical it is. According to Fagerberg, continuous improvement or marginal innovation was viewed by Schumpeter as less important than radical or revolutionary technological change (Fagerberg, 2003). This concept of innovation as creative destruction is still popular today and is associated with the idea of disruptive technological change: the mobile phone making landlines redundant, electronic media leading to the demise of print, and so on. There is some doubt, however, as to how important radical innovation really is in terms of its economic impact. Using electric power generation as an example, in the 50-year period between 1910 and 1960, there was an 88 per cent reduction in the amount of coal needed to produce 1 kilowatt-hour of electricity. This change, which has had a huge economic impact, was due to a 'stream of minor plant improvements', including a rise in operating temperatures and pressures made possible by new alloy steels, more sophisticated boiler design, improvements in turbine efficiency, and 'the addition of components such as feed-water heaters and stack economizers' (Kline and Rosenberg, 1986: 283). This has led some to argue that the bulk of economic benefits comes from incremental innovations and improvements (for example, Lundvall et al., 1992; Fagerberg, 2003).

Schumpeter's idea of the heroic risk-taking entrepreneur being central to innovation remains heavily embedded in today's narratives (one only has to listen to a TEDx talk online or consider the lionization of individuals such as Bill Gates and Steve Jobs). Those narratives have, in turn, tended towards an unbalanced and unhelpful emphasis in technology innovation policy on two instruments that entrepreneurs bring to bear in the innovation process – private capital and intellectual property rights – while paying less attention to the many other drivers that influence the process, something that will be explored in later chapters.

Despite this, there are counternarratives that question exactly how important the role of the entrepreneur is. Some of these relate directly to views on the relative importance of the private sector versus the state (see the later discussion of Mazzucato's work on the 'entrepreneurial state'). Others relate to the idea that innovation is a collective achievement and occurs in systems.

The innovation systems approach

Early views of technological innovation saw it as a linear process of scientific research producing new ideas, followed by development of those ideas into practical applications that would then go into production and be marketed (Kline and Rosenberg, 1986: 286). Critiques of this view mentioned:

- the lack of feedback loops in the model to allow for the fact that scientific research can itself be driven by perceived technological needs (Kline and Rosenberg, 1986);
- the fact that, although some technological innovation stems from advances in science, much is driven through 'learning by doing' as using and interacting with technology leads to incremental improvements (Lundvall, 2004);
- the problems with assuming the innovation process is contained within the 'firm' or commercial enterprise, whereas, in reality, knowledge and motivation can be influenced by a much wider set of factors, including movement of individuals from one company or sector to another, and published research.

The idea that technological innovation can be the result of a network of interactions and feedback loops among a variety of actors, as opposed to a linear process driven solely by 'the entrepreneur', has led to the development of a body of work around innovation systems and an interest in polices that can support the strengthening of national innovation systems (NIS) (Lundvall, 2004).

Andrew Watkins and colleagues provide a good overview of the development of innovation systems approaches (Watkins et al., 2014). According to their analysis, early work on NIS centred on companies as the principal unit through which innovation is developed. Companies interacted with universities, which provided new ideas and talent for their research and development departments. Governments played a 'necessary but almost passive' role providing support through incentives and regulatory environments. Politics, the process by which governments are both informed and exert influence, was largely omitted from the analysis. In the mid-1990s a strand of the literature started to explore the role of the intermediary institutions, such as research councils, funding bodies, and universities, that facilitate knowledge flows between innovators and policymakers. By the early 2000s, research had expanded to include the roles of industry associations, and a broader and more active role for government was acknowledged (although, as will be shown later, it is questionable how much that recognition has actually penetrated policy in practice). There was also some experimentation to see if the nation state was the best vehicle for analysis or whether subnational systems, technology systems, or functional processes were more useful. All approaches provided some interesting insights but, given the role national governments have in setting policy environments, the national innovation system has proven an enduring analytical framework.

Mapping of innovation systems has been done at various levels of complexity to show the actors involved, the interrelationships between them, and the factors governing the outcomes of those interrelations. A UK government report assessing the relative strength of its NIS, for example, includes a map of the UK innovation system with more than 100 components and their associated feedback loops, which is too complex to reproduce here (Allas, 2014). Alternative and simpler representations of innovation systems have been produced, including one by the US state of Connecticut, which has been adapted several times in the literature to portray the generic concept of an innovation system (see Figure 7.3).

Innovation systems approaches can go beyond theoretical analysis of how innovation happens to practical attempts to optimize the conditions for innovation to take place. The analysis of the UK NIS, for example, led the Department of Business Innovation and Skills to assess its main strengths and weaknesses relative to the systems of the competitor economies of Australia, Canada, Finland, France, Germany, Japan, South Korea, and the United States. The results of this analysis were then used to make recommendations for further research and to provide initial pointers for future government policy (for example, the need to invest further in basic numeracy, literacy, ICTs, problem-solving and science, technology, engineering, and mathematics skills in the education system, or the need for further research into why the UK's public and private investment in research and development is low relative to its competitors).

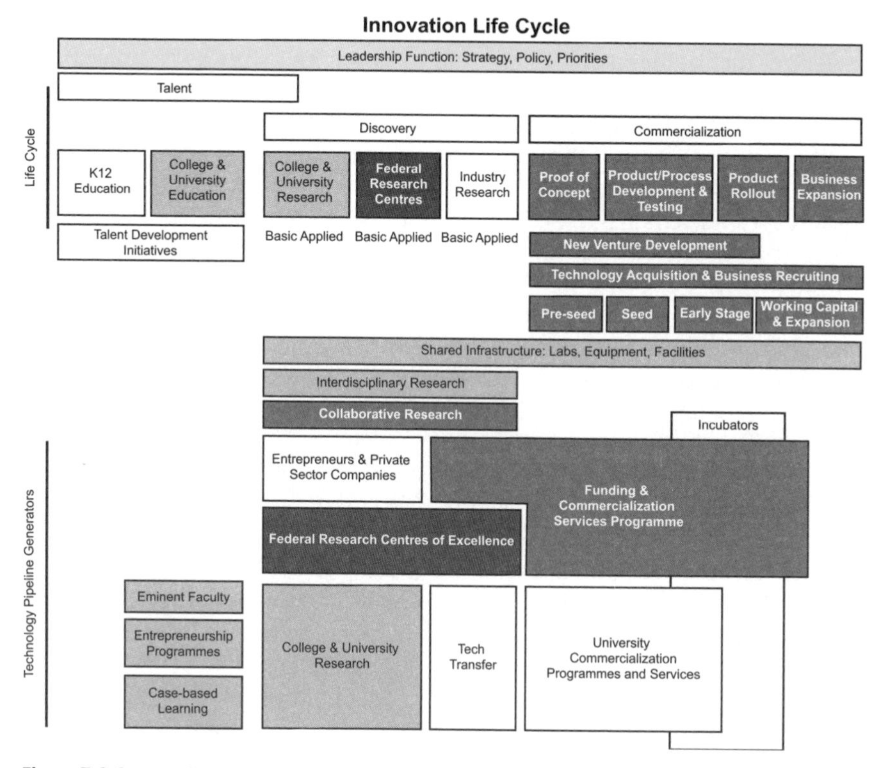

Figure 7.3 Connecticut's innovation system

Source: Abromaitis et al., 2006

From a Technology Justice perspective, innovation systems thinking is important as it reveals the multiplicity of actors that influence innovation processes and the need for a systemic approach to the crafting of policy if a fair space for innovation is to be created. The role of government in national innovation systems emerges as an important area of debate, particularly given the urgent need to find a response to climate change and to establish a safe, inclusive, and sustainable space for development. That debate – which is generally posed as a choice between government as a facilitator setting the 'rules of the game' and then stepping back to let the market work versus government as an active player setting priorities, making investments, and taking risks itself – is explored further in Chapter 10 (Trace, 2016).

Innovation systems thinking has also been important in changing views on how developing countries might use technology innovation and technology transfer to speed up economic development and lift living standards. This is examined in the next section.

Innovation systems and developing economies – insights and problems

Thinking and practice with respect to how to promote technology transfer or technology 'catch-up' in developing country economies has, unsurprisingly, reflected the changing ideas on how technology innovation impacts on economic growth outlined earlier. In the 1960s and 1970s, a common version of neoclassical growth theory, influenced by the likes of Solow, meant that technology (or at least technical knowledge) was conceived as a 'public good' freely available to anyone. Moreover, it was recognized that the nature of technology was changing. Economists such as Veblen had argued as early as 1915[2] that, in the past, technology had, in effect, been tacit or bound up in the knowledge and skills of individuals and thus technology transfer required the physical migration of skilled workers from one location to another. With the advent of machine technology, technical knowledge was essentially embodied or codified in hardware, which could be easily transferred anywhere. This combination of ideas of technology as a public good and something easily embodied in machines led to the conclusion that technology transfer should therefore happen automatically. Once someone had put in the effort and investment necessary to develop a new and successful technology, others would naturally adopt that technology 'ready-made' rather than invest in the expense of developing their own alternative. Over time, the argument went, economies would naturally converge as technology developed in the high-income countries was adopted by others in this manner, provided the market was left alone to do its work (Fagerberg et al., 2009).

In development policy terms this was the era of direct technology transfer as the proposed solution to technology catch-up. In simplistic terms, if the problem was agricultural production then the solution was to import modern machinery – tractors, combine harvesters, and so on – and the latest Green Revolution technologies – hybrid seeds, fertilizers, pesticides, and herbicides. But there was plenty of evidence in the 1960s and 1970s that simply transplanting pieces of equipment from the developed to the developing world wasn't the whole answer. Agricultural machinery that could not be maintained, was too expensive to operate, or was unsuitable for local conditions was abandoned, and large swathes of Africa remained untouched by the Green Revolution, to give but two examples.

For economists, two pieces of evidence in particular suggested that the idea of technology as a freely available public good could not explain how technology innovation occurred and was spread in less developed economies. Firstly, as time went by, it became obvious that developing and developed world economies were, in many cases, not converging (as the public good theory of technology predicted) but diverging. Secondly, some of the most well-known examples of countries moving out of poverty and making rapid strides in raising living standards – South Korea, Taiwan, and Singapore – followed a route that was anything but passively waiting for the market to do its job (Fagerberg et al., 2009). These countries placed great emphasis on building indigenous technical capabilities. Korean academic Linsu Kim produced a number of research papers from 1980 onwards developing an analytical framework to describe how the development of technological capability influenced industrialization in Korea (for example, Kim, 1980, 1997, 2000). In these papers he identified three stages of Korea's technological capacity development: the duplicative imitation stage, the creative imitation stage, and the innovation stage. He also highlighted a number of proactive actions taken by the Korean government and other institutions that underpinned the country's move through those stages and its journey from one of the poorest nations in the world to a high-technology developed economy in just three decades.[3] These included:

- promoting export as a policy instrument that forced Korea's companies to compete against the international market (while protecting the domestic market from foreign competition in its infancy);
- expanding the quality of education at all levels to build a knowledge base;
- having a liberal policy initially on brain drain to allow Korean scientists to gain experience and further develop their skills overseas when the national market for their skills was immature, followed by an active repatriation effort when expansion was underway;
- using the recruitment of high-calibre human resources, foreign technology transfer, and in-house research to build a high tacit knowledge base in the labour force;
- evolving technology transfer strategies over time, starting with reverse engineering of more mature and simple technologies, then also using informal mechanisms such as the reverse brain drain or the manufacturing of original equipment on behalf of foreign suppliers, and, as technology became more complex, using foreign licensing agreements, foreign direct investment, and, ultimately, indigenous research and development capability;

- the government setting ambitious goals that, in effect, created crises that led to high-intensity innovation efforts by companies to resolve them;
- investing in government research institutes and evolving their role over time to meet changing needs.

This analysis is essentially an innovation systems approach to looking at the development of technology and technical capability in Korea and reflects where much current thinking lies. Technological catch-up in developing economies is *not* automatic; deliberate action is required to support the acquisition of technical capabilities and this is achieved not by the entrepreneur in isolation, but through myriad interactions between different national and international actors, including deliberate shaping and directional efforts by government.

Despite the Korean example, questions have been raised regarding the relevance of the NIS approach to developing economies, given that it is built largely on developed-country economic analysis. Arocena and Sutz, for example, argued back in 2001 that the concept might be less relevant to Latin America, suggesting:

> that the types of growth actually prevailing in Latin America are based on the intensive and frequently damaging use of natural resources and/or in assembling activities ... as well as in low salaries and weak social and environmental regulations. In most cases, knowledge, innovation and advanced learning play a marginal role. ... In such context, Innovation Systems look more fragmented than systemic, show a low density of national innovative relations, and depend essentially on innovation coming from abroad. (Arocena and Sutz, 2001: 11)

More recently, Watkins and colleagues suggested that the application of NIS thinking to developing countries was also complicated by the implication that an innovation system was something that 'continually, and often rapidly correct(s) ... inefficient pathways toward the advancement and maturing of industries' and that 'the emphasis is on ... high tech industries' (Watkins et al., 2014: 7). They argue that, because of this, applying the approach in developing countries may very well 'miss existing innovation systems in these countries that are based on more slowly developing, less technology-driven industries (for example, agriculture and craft industries) – possibly hindering the application of the NIS concept as a development tool and strategy' (Watkins et al., 2014: 7).

An alternative view maintains that the NIS approach is useful in developing countries but must take into account that their innovation systems are often substantially different from the mature innovation systems found in developed economies and that, 'just imitating innovation policies practiced in developed countries is unlikely to deliver the expected results' (Chaminade et al., 2009: 2). In less developed economies, the argument goes, innovation systems should be thought of as fragmentary and evolving, where only some of the building blocks are in place and interactions between the elements are still being formed (see Figure 7.4). This has important implications for the identification of system constraints as the question shifts from 'which

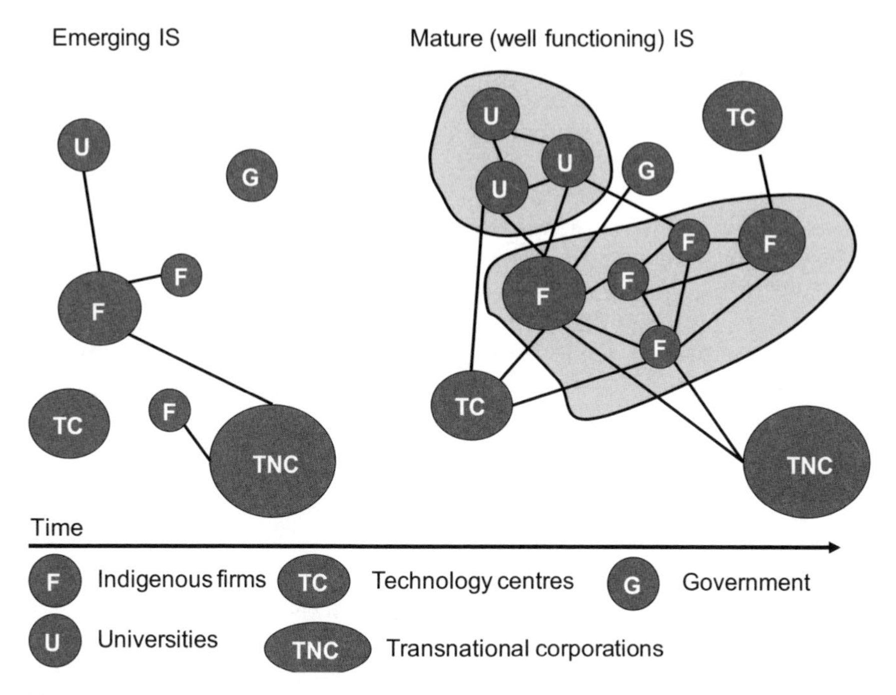

Figure 7.4 Stages in the development of an innovation system
Source: Chaminade et al., 2006: 30

elements and relationships within the system are weak?' to 'what elements are critical for the emergence and development of ... a fully-fledged socially inclusive innovation system and how [can] systemic innovation policies ... be designed?' (Chaminade et al., 2009: 8).

Meanwhile, the Technology Executive Committee of the United Nations Framework Convention on Climate Change (TEC UNFCC) argues for the relevance of NIS thinking to developing countries but suggests that, given the enormity of the task of strengthening entire emerging and fragmented national systems, developing countries should instead consider beginning by strengthening just part of a system to meet specific climate and development challenges. 'In this way, a country can efficiently allocate its resources to strengthen innovation system elements that are most relevant to successful implementation of its priority technological pathways in the climate arena' (TEC UNFCC, 2015: 7).

What is noticeable in the literature is the lack of active or recent research on the actual application of NIS approaches in developing countries. Out of the 98 references cited in the literature review by Watkins et al. mentioned earlier, just five are papers written since 2010, and all of them focus on either the emerging economies of Brazil, India, and China or Asian 'tiger' economies, such as Taiwan. Two-thirds of the references cited were more than 10 years old, something that seems common to all the literature reviewed for this chapter. It is noted that 'there is little information on the current state of developing country NIS's' (TEC UNFCC, 2015: 2) and more research is required; in particular, 'identifying which elements and relationships are critical in emergent systems of innovation requires a deeper analysis of the specificities of systems of innovation in developing countries' (Chaminade et al., 2009: 8).

As mentioned in the introduction of this book, both the Sustainable Development Goals and the United Nations Framework Convention on Climate Change processes have put considerable emphasis on the importance of technology transfer to the developing world to meet poverty and climate goals. The scale of investments envisaged is huge. The World Bank has estimated the requirement for developing-country climate-change mitigation finance alone, much of which would be around energy technology transfer, to be between $265 and $565 bn per annum for the next 20 years (Montes, 2013). In this context, the absence of research on existing innovation systems in developing countries and how they can be strengthened to support the envisaged level of technology transfer is a massive knowledge gap that needs to be addressed urgently.

Notes

1. As Mariana Mazzucato, RM Philips Professor in the Economics of Innovation at the University of Sussex, has wryly pointed out: 'if the underlying model was found so deficient that it could not explain 90 per cent of the dependent variable it was describing then it should have been thrown out and a new model developed' (Mazzucato, 2013: 33).
2. In his study *Imperial Germany and the Industrial Revolution* (1915), as cited by Fagerberg et al., 2009.
3. Kim also noted that some of these, notably protection of the domestic market in the early stages of developing technical capabilities and reverse engineering of products, would not be routes open to developing countries today because of the World Trade Organization rules and more aggressive protection of intellectual property rights which have evolved in recent decades.

References

Abromaitis, J. et al. (2006) *Connecticut's Innovation Network,* Connecticut: Department of Economic and Community Development.

Allas, T. (2014) *Insights from International Benchmarking of the UK Science and Innovation System,* BIS Analysis Paper No. 3, Department for Business, Innovation and Skills, London: UK Government.

Arocena, R. and Sutz, J. (2001) 'Innovation systems and developing countries', Working Paper no. 02-05, Copenhagen: Danish Research Unit for Industrial Dynamics <http://www3.druid.dk/wp/20020005.pdf>.

Chaminade, C. and Vang, J., (2006) *Globalisation of Knowledge Production and Regional Innovation Policy: Supporting Specialized Hubs in Developing Countries,* CIRCLE Working Paper 2006/15, Lund: Lund University Sweden. <http://wp.circle.lu.se/upload/CIRCLE/workingpapers/200615_Charminade_Vang.pdf>.

Later published in

C. Chaminade, Lundvall, B.-A., Joseph, K.J. and Vang, J., (2009) 'Designing IS policies for development in the new global context' in Lundvall, B.A et al *Innovation Systems and Developing Countries – Building Domestic Capabilities in a Global Setting*. Cheltenham: Edward Elgar. (Figure 13.1, page 366)

Fagerberg, J. (2003) 'Innovation: a guide to the literature', in J. Fagerberg and D.C. Mowery (eds), *The Oxford Handbook of Innovation,* Oxford: Oxford University Press <http://dx.doi.org/10.1093/oxfordhb/9780199286805.003.0001>.

Fagerberg, J., Srholec, M. and Verspagen, B. (2009) 'Innovation and economic development', UNU-Merit Working Paper Series, Maastricht Economic and Social Research and Training Centre on Innovation and Technology, Maastricht: United Nations University.

Kim, L. (1980) 'Stages of development of industrial technology in a developing country: a model', *Research Policy*, 9: 254–77 <http://dx.doi.org/10.1016/0048-7333(80)90003-7>.

Kim, L. (1997) 'Analytical frameworks', in L. Kim (ed.), *Imitation to innovation – The Dynamics of Korea's Technological Learning*, pp. 85–105, Boston, MA: Harvard Business School Press.

Kim, L. (2000) 'The dynamics of technological learning in industrialisation', INTECH Discussion Paper Series, Maastricht: United Nations University.

Kline, S.J. and Rosenberg, N. (1986) 'An overview of innovation', in R. Landau and N. Rosenberg (eds), *The Positive Sum Strategy: Harnessing Technology for Economic Growth*, pp. 275–305, Washington, DC: National Academy Press.

Lundvall, B. et al. (1992). *National Systems of innovation: Towards a theory of innovation and interactive learning* . London: Pinter.

Lundvall, B. (2004) 'National innovation systems – analytical concept and development tool', presented at the *DRUID Tenth Anniversary Summer Conference on Dynamics of Industry and Innovation: Organizations, Networks and Systems*, Copenhagen: Aalborg University.

Mazzucato, M. (2013) *The Entrepreneurial State: Debunking Public vs. Private Sector Myths*, London: Anthem Press.

Montes, M.F. (2013) 'Climate change financing requirements of developing countries', *Climate Policy Brief 11*, Geneva: South Centre.

Schumpeter, M. (2008) [1911] *The Theory of Economic Development: An Inquiry into Profits, Capital, Credit, Interest and the Business Cycle*, trans. R. Opie, New Brunswick: Transaction Publishers.

Śledzik, K. (2013) 'Schumpeter's view on innovation and entrepreneurship', in S. Hittmar (ed.), *Management Trends in Theory and Practice*, pp. 89–95, University of Zilina <http://dx.doi.org/10.2139/ssrn.2257783>.

Solow, R.M. (1987) 'Growth theory and after', Prize Lecture, *Nobel Prize*, <http://www.nobelprize.org/nobel_prizes/economic-sciences/laureates/1987/solow-lecture.html> [accessed 11 October 2015].

Solow, R. (2007, 17 May) 'Robert Solow on Joseph Schumpeter', 17 May, *Economist's View*, <http://economistsview.typepad.com/economistsview/2007/05/robert_solow_on.html> [accessed 11 September 2015].

TEC UNFCCC (2015) 'Strengthening national innovation systems to enhance action on climate change', TEC Brief #7, Bonn: Technology Executive Committee, United Nations Framework Convention on Climate Change.

critical review of the literature', IKD Working Paper no. 70, Milton Keynes: The Open University.

Ziegler, R. (2015) 'Justice and innovation – towards principles for creating a fair space for innovation', *Journal of Responsible Innovation*, 2(2): 184–200 <http://dx.doi.org/10.1080/23299460.2015.1057796>.

CHAPTER 8

Technology Justice and innovation systems in practice

Justice and the management of risk in innovation

Science and technology innovation carries with it risks. In Chapter 1, technological determinism was shown to be a faulty concept; it is not possible to anticipate with certainty where a particular line of research will lead, how it will be used, or what its impact will be. Einstein did not predict the atomic bomb when he arrived at the formula $E=mc^2$. Emerging technologies can be based on wide and complex fields of immature scientific research with little history to identify risks and anticipate consequences. Risto Karinen and David Guston writing on nanotechnology, for example, note that: 'because nanotechnologies are currently inchoate, even those stakeholders who recognise an interest in them often operate with only loosely formed and ill-conceived expectations of them' (Karinen and Guston, 2010).

Anticipating and mitigating as yet unknown adverse impacts of a technology innovation process is a significant risk-management challenge in itself. The following section looks at the precautionary principle as an approach to the management of risk in technology innovation processes.

The consequence for society or the environment of backing the wrong innovation has the potential to be catastrophic. Chapter 5 used the links between fossil fuels and climate change, and between industrialized agriculture and biodiversity loss to illustrate this point. But negative impacts can occur at the meso as well as the macro scale, as shown by a European Environment Agency report that reviews a range of unpredicted and negative outcomes from technological innovations (mostly the development of chemical compounds) over the last century, a selection of which are summarized in Table 8.1.

In the absence of certainty and an ability to accurately predict the risk of negative impacts, the governance of innovation becomes a difficult and contested space. The examples given in Table 8.1 are, almost without exception, stories of conflicting views. Industry generally insists on the safety of its product and argues against regulation it views as based on inconclusive scientific research or misguided public concerns. In some cases, governments and regulatory authorities err on the side of public concern (as was the case with imidacloprid and honey bees) and sometimes on the side of industry (as was the case for many years in Japan with the discharge of methylmercury into the sea and in the US with minimum exposure levels for beryllium).

In all the cases listed in Table 8.1, there were early warnings of potential risks to human health or to the broader environment and opportunities to

http://dx.doi.org/10.3362/9781780449043.009

Table 8.1 Technologies introduced in the last century and their unexpected negative impacts

Technology innovation	*Impact*
Lead in petrol (1925–2012)	Neurotoxin causing brain damage, especially in young children
Perchlorethylene (PCE), used in the production of plastic linings for drinking-water distribution pipes in the US (late 1960s–70s)	Carcinogenic agent leaked into water supply in New England
Methylmercury used in the production of acetaldehyde (1932–68)	Mercury poisoning outbreak in Japan as a result of factory effluent flowing into the sea, causing mercury to enter the food chain via contaminated fish
Beryllium, used in manufacturing processes, including the production of nuclear weapons (1940s–to date)	Chronic beryllium disease, an irreversible inflammatory lung disease contracted by workers exposed to the metal
Dibromochloropropane (DBCP), a pesticide (1955–90s)	Deficient or absent sperm count in agricultural workers in the US and Latin America
Seed-dressing systemic insecticides containing imidacloprid (1994–2004)	Collapse of honey bee colonies in France

Source: abstracted from European Environment Agency, 2013

avoid damage were missed (Appendix 1 provides details of the nature of those warning signs and industry and government's response in each case). There is a wide set of case studies demonstrating how early warnings of danger were ignored, resulting in late action and considerable damage (European Environment Agency, 2001; Gee, 2013). Some of these studies address chemicals or other substances that are now widely known to be hazardous, including (in addition to those shown in Table 8.1) asbestos, benzene, bovine spongiform encephalopathy (BSE or mad cow disease), diethylstilboestrol, tributyl tin, polychlorinated biphenyls, dichlorodiphenyltrichloroethane (DDT), vinyl chloride monomer, and booster biocides. David Gee argues that, combined, these cases illustrate the need for a more precautionary approach to be applied to emerging chemical risks, such as:

- bisphenol A (BPA), a very common chemical used to make the polycarbonate plastics used in everything from household electronics to food containers and babies' bottles. BPA is known to mimic the female hormone oestrogen and has been found to leach from the materials where it is used;
- nicotinoid pesticides, where it has been shown that in some circumstances the application of authorized doses of insecticide can affect wild pollinators such as honey bees as well as earthworms and beneficial biological control agents;
- endocrine-disrupting substances which are present in some consumer products, including pharmaceuticals, such as ethinyl oestrodiol in the pregnancy pill (which is linked to the feminization of fish) (Gee, 2013).

Gee further suggests that:

> the histories of well-known technologies, such as X-rays, fishing techniques, fossil fuel power sources and early nuclear plants, can also provide lessons for prudent action on the potential hazards of such emerging technologies as nanotechnology, genetically modified food, radio-frequency from mobile phones, and the new generation of nuclear plants. (Gee, 2013: 644)

Proposals to trial geoengineering[1] techniques to curb the effects of global warming is another example of emerging technologies around which the need to apply the precautionary approach provokes much discussion (see, for example, Elliot, 2010; Powell et al., 2010; Tedsen and Homann, 2013).

That a precautionary principle should be applied to science and technology innovation to mitigate and manage risk has widespread theoretical acceptance, as evidenced by Gee's list of 13 international treaties that contain elements of the precautionary principle (see Box 8.1). But the lack of a consistent definition of the principle across these treaties illustrates the level of confusion that exists around its application. This has led to a variety of criticisms of the principle itself, including that it is often applied using solely scientific criteria and ignores ethical concerns in decision-making (Ahteensuu, 2007); that weak forms of the precautionary principle are tautological (worded in such a way that the application of the precautionary principle itself could be seen as a risk to avoid) (Mandel and Gathii, 2006); that it has been used as a pretext for erecting protectionist trade barriers (for example, the impact of EU bans on genetically modified ingredients in food and growth hormones in meat on the US) (Lofstedt, 2004); and that it can block the flow of benefits from scientific advance[2] (Paris Tech Review, 2014). The latter idea is often voiced as a fear that the precautionary principle, if applied, will result in too many 'false positives' – incidents where either minor or even non-existent risks are regulated because of ill-informed or unwarranted public concerns, leading to a stagnation of research. This fear doesn't seem to have much foundation in fact, though. A review of 88 alleged false positives from studies in the European Union, for example, found most to be either real risks, cases where 'the jury is still out', unregulated alarms, or risk-risk trade-offs, with only four genuine 'false positives'. This led the reviewers to conclude that:

> the fear of false positives is misplaced and should not be a rationale for avoiding precautionary actions where warranted. False positives are few and far between as compared to false negatives and carefully designed precautionary actions can stimulate innovation, even if the risk turns out not to be real or as serious as initially feared. (Hansen and Tickner, 2013)

Gee cites the European Environment Agency's own formulation of the precautionary principle as an attempt to address the confusion around definition and interpretation:

Box 8.1 International treaties that contain references to the application of the precautionary principle

1. Rio Declaration on Environment and Development, 1992
2. European Union's Treaty on the Functioning of the EU, Article 191(2)
3. Regulation (EC) No 1907/2006 concerning the Registration, Evaluation, Authorisation and Restriction of Chemicals (REACH)
4. UN Framework Convention on Climate Change, 1992
5. EU Directive 2001/18/EC on deliberate release of GMOs, Article 4(1)
6. Cartegena Protocol on Biosafety, 2000, Article 11(10)
7. Regulation (EC) No 178/2002 establishing the European Food Safety Authority and procedures in matters of food safety, Article 7
8. EU Regulation 1107/2009 on plant protection products, Article 1(4)
9. London International Maritime Organisation Convention on the Control of Harmful Anti-fouling Systems on Ships, 2000, Articles 6(3) and (4)
10. European Court of Justice in the BSE case (Case C-157/96, National Farmers Union and others, 1998, ECR 1-2211)
11. Stockholm Convention on Persistent Organic Pollutants, 2001
12. WTO Agreement on Sanitary and Phytosanitary Measures (SPS Agreement), Article 5(7)
13. European Commission communication on the precautionary principle, 2 February 2000

> The precautionary principle provides justification for public policy and other actions in situations of *scientific complexity, uncertainty and ignorance,* where there may be a need to act in order to avoid, or reduce, potentially serious or irreversible threats to health and/or the environment, using an *appropriate strength of scientific evidence,* and taking into account *the pros and cons of action and inaction* and their distribution. (Gee, 2013: 649).

He goes on to offer useful definitions of some common concepts in the precautionary principle and advice on their use in practice (see Table 8.2).

Of course, objectors to the precautionary principle are not only concerned with the confusion surrounding its definition. There is also a strong and persistent level of opposition to the use of the precautionary principle from powerful corporations on commercial grounds, as well as from some scientists who have been compromised by sources of funding, and from politicians fearing high economic or political costs arising from its application.

In the opposite camp there is criticism of a lack of research funding for looking at risk. Gee claims that 'public research funding by the EU on nanotechnology, biotechnology and information technology was heavily biased towards product development, with only about 3 per cent of the € 28.5 bn budget spent on investigating their potential hazards' (Gee, 2013: 646). Other authors document a similar imbalance on research into genetic modification in the US: 'in the 11-year period of 1992 to 2002,

Table 8.2 Common concepts used in debates on the precautionary principle and appropriate actions

Situation	*Nature of knowledge*	*Type of action taken*
Risk	**'Known' impacts** and **'known' probabilities** e.g. regarding *asbestos* from 1930	**Prevention:** action to reduce known hazards e.g. eliminating exposure to asbestos dust
Uncertainty[1]	**'Likely' impacts** but **'unknown' probabilities** e.g. regarding antibiotics in animal feed and associated human resistance to those antibiotics, from 1965	**Precaution:** action taken to reduce exposure to plausible hazards e.g. the EU ban on antibiotic growth promoters in 1999
Ignorance	**'Unknown' impacts** and therefore **'unknown' probabilities** e.g. the then unknown but later 'surprises' of the ozone layer 'hole' from CFCs, pre-1974; the mesothelioma cancer from asbestos pre-1959; the rate of Greenland ice sheet melting pre-2007	**Precaution:** action taken to anticipate, identify earlier, and reduce the extent and impact of 'surprises' e.g. by using intrinsic properties of chemicals e.g. persistence, bioaccumulation, spatial range; using analogies; long-term monitoring; and using robust, diverse, and adaptable technologies that can help minimize impacts of 'surprises'
Ambiguity	Concerning the different **values and interpretations** about information used by stakeholders e.g. in invasive alien species cases where a species can be welcomed by some but not others	**Participatory precaution:** stakeholder engagement in decision-making about innovations and their potential hazards
Variability	The natural differences in population or ecosystem exposures and sensitivities to harmful agents	**Obtain more information** in order to minimize simplistic assumptions about average exposures and sensitivities
Indeterminacy	Unpredictable uses of technologies e.g. use of X-rays in children's shoe shops in the 1950s	**Pre-market benefit assessment** of novel uses of a technology with potential hazards

1. different types, sources, and levels of uncertainty can be identified (citing Walker, 2003)
Source: Gee, 2013: 656

the USDA spent approximately $1.8 bn on biotechnology research and approximately $18 m on risk-related research' (Mellon and Rissler, 2003).

The selection of false negatives in Table 8.1, examples of early warnings of risks that were ignored, shows the high social, economic, and environmental costs of failing to implement a precautionary principle in the governance of emerging technologies. Creating consensus around what the application of the precautionary principle would look like in practice, building on the concepts in Table 8.2, is clearly important to better manage the risks inherent in science and technology research and innovation.

As a tool to create Technology Justice and a 'fair space' for technology innovation, the precautionary principle has its limitations, however, even if there were more of a consensus on its purpose and use. Primarily intended as a tool to manage unforeseen risks arising from the innovation process, it says nothing about the purpose of research or whether, to refer back to Ziegler's ideas from Chapter 7, innovation is directed in a way that ensures it contributes to the long-term stability of society (by finding creative responses to societal challenges, such as climate change), or is specifically aimed at improving conditions for the least advantaged in the present. This is the subject of the remainder of this chapter.

Justice and the shaping of the purpose of innovation

There is indeed a need for a more comprehensive governance approach that not only seeks to manage risk but also sets the direction of research and development in a way that provides socially and environmentally useful outcomes. The argument to be made here is that the current drivers of innovation fail to deliver research and development programmes that adequately address the big 'fair space' questions of environmental sustainability and poverty. Two examples will be used to illustrate this: technology innovation in the fields of health and agriculture.

Technology innovation for health – whose priorities?

In 1990 a mismatch was identified between the health research and development (R&D) that is actually undertaken and that which is needed (Commission on Health Research for Development, 1990). At that time, it was demonstrated that less than 10 per cent of global health research expenditure was spent on the health problems of developing countries, despite the fact that the developing world contained more than 90 per cent of the world's burden of preventable deaths. This imbalance became known as the '10/90-gap'.

The nature of the 10/90-gap has changed considerably since 1990. Global research funding for health has increased eightfold from $30 bn a year in 1986 to $246 bn in 2010 (Viergever, 2013) and the distribution of the global disease burden is different, with an increasing proportion of the burden in all regions but sub-Saharan Africa being taken up by non-communicable diseases (GBD 2013 Risk Factors Collaborators, 2015). A variety of new approaches to innovation have been suggested and tested, such as advanced market commitments and a Global Health Observatory (see later), to encourage action on previously neglected areas of health research.

Having said all that, a substantial disparity remains in the resources applied to research into the disease burdens of higher versus lower and middle-income countries. For example, a 2013 paper in the *Lancet* notes that:

> Of the $214 billion invested in high-income countries, 60 per cent of health R&D investments came from the business sector, 30 per cent from the public sector, and about 10 per cent from other sources (including private non-profit organisations). Only about 1 per cent of all health R&D investments were allocated to neglected diseases in 2010. Diseases of relevance to

> high-income countries were investigated in clinical trials seven-to-eight-times more often than were diseases whose burden lies mainly in low-income and middle-income countries. (Røttingen et al., 2013)

The 10/90-gap is part of a wider problem of neglected groups. This neglect can be seen in the lack of development of new medicines that are affordable for all as well as the relatively low amount of R&D into diseases that mainly affect developing countries. But the problem of the gap goes beyond the developing world and involves, for example, the lack of research into new antibiotics, or into so-called orphan diseases which may have very significant impacts but on only a relatively small population. In addition to neglected groups of people, there are neglected products. Research is generally focused on the development of drugs and vaccines more so than on the development of diagnostic tools or what are known as platform technologies (technologies that can potentially be applied to more than one disease).

One reason for the continuation of the 10/90-gap is the dependency of pharmaceutical research funding on market forces. As noted, about 60 per cent of all health R&D funding comes from the private sector and is therefore focused on products that will create the best financial return as opposed to necessarily meeting the greatest need. This not only steers research to markets with more buying power in the developed world, but also, within that market, tends to deliver 'innovation' that is of only marginal value: the 'me too' drugs that offer little additional therapeutic value but provide a good financial return because their development costs are relatively low. In the US, in the period 1993 to 2004, only 357 of the 1,072 drugs approved by the Food and Drug Administration (FDA) were classified as new molecular entities (NMEs) or completely new medicines. The remainder were minor variants on existing medicines (the 'me too' drugs), for example the same drug repackaged in different dosages. Within the relatively small number of NMEs produced, even fewer were seen as important advances and given a priority rating by the FDA – just 146 or around 14 per cent of the total number of drugs approved (Mazzucato, 2013: 66). This problem was recognized by Bill Gates who, in an interview about his foundation's work on malaria, bewailed the fact that: 'Our priorities are tilted by marketplace imperatives. The malaria vaccine in humanist terms is the biggest need. But it gets virtually no funding. But if you are working on male baldness ... you get an order of magnitude more research funding' (quoted in Solon, 2013).

The scale of the problem of relying on market-driven innovation to deliver products for diseases that primarily affect the poor is clear when you look at the levels of investment broken down by sector. Almost 90 per cent of the global health R&D spend occurs in high-income countries (around $214 bn in 2010). The annual spend on neglected diseases of poverty is just over 1 per cent of that figure – just $2.5 bn. Although the private sector provides the majority of global health R&D funds, its share of the very small proportion of funding allocated to neglected diseases is disproportionately low, at just over 16 per cent (see Figure 8.1). The bulk of R&D on diseases that predominantly affect the developing world is carried out using public funds with some additional support from philanthropic institutions (principally the Gates Foundation and the Wellcome Trust).

Market mechanisms are failing to deliver products that are directed towards the health challenges of the developing world or which are affordable by the majority of the population living there. Furthermore, by their nature, competitive markets are not really supportive of open innovation which, in itself, acts as an impediment to the efficiency of an already overburdened R&D system. Indeed, commercial pressures have been seen to encourage companies to misrepresent or exclude data from public reports that would question the relative efficacy of their products. For example, in the 1990s, the company Pfizer created a new drug for treating clinical depression called Reboxetine. By selectively reporting the results of just some of the clinical trials carried out on the drug, it published data that suggested the efficacy of the drug was much better compared to rival compounds than was actually the case (*The Economist,* 2015).

Public and philanthropic funding does not necessarily fare better in terms of delivering medical innovations focused on the needs of the developing world, despite the fact that it accounts for the remaining 40 per cent of global health R&D spend and 80 per cent of the R&D spend on neglected diseases (Viergever, 2013: 5). An evaluation of the research focus of the 25 UK universities receiving the highest amount of funding in 2010–11 from

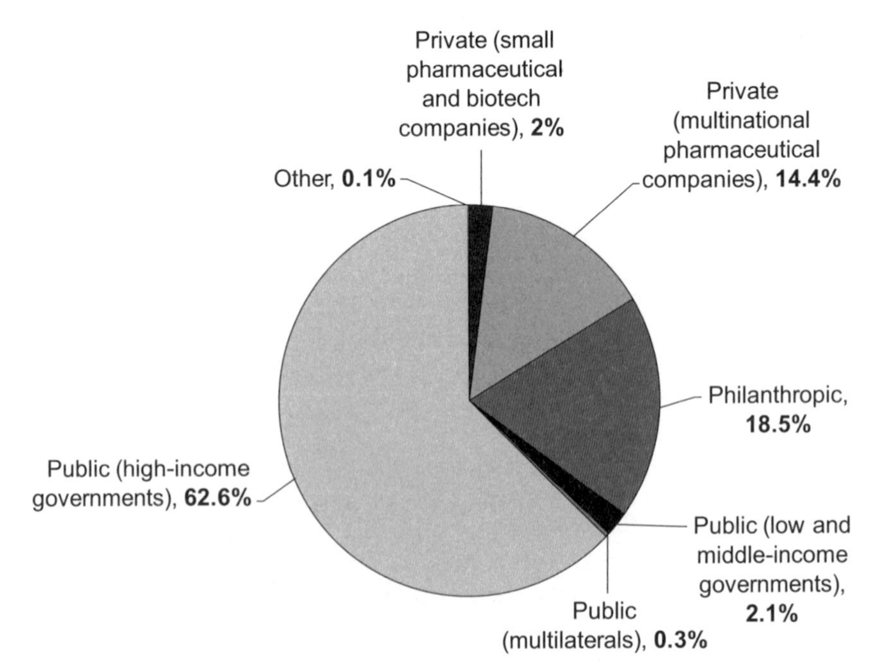

Figure 8.1 Sources of funding for R&D on neglected diseases, 2009
Source: WHO, 2012a: 126

the primary public funding agency in the UK, the Medical Research Council, provides an interesting insight (Universities Allied for Essential Medicines UK, 2014). The evaluators argued that, because universities are major drivers of innovation, with academic laboratories being responsible for 25–33 per cent of all new medicines, it is important to understand whether the public funding invested in these endeavours is being used to the greatest effect. The study used information in the public domain and from questionnaires administered to the universities to evaluate two questions:

- Are universities investing in global health research and medical research that addresses the neglected health disease needs of low-income communities worldwide?
- Do universities share new discoveries in ways that ensure treatments reach people in developing countries at affordable prices, and how much of their research is freely available online? (Universities Allied for Essential Medicines UK, 2014)

The study used a group of 11 metrics to answer these questions and graded a combined weighted overall score for each university on a scale from A+ (a strong yes to both questions) to D (very little evidence to support a positive answer to either question).[3] Expectations were not set very high. To maintain the highest score, for example, a university would have to spend just 8 per cent of its health research budget on neglected diseases[4] and 8 per cent on projects devoted to health in low and lower–middle-income countries. Of the 25 universities reviewed, only six (24 per cent) scored higher than a C+. The same exercise repeated on the 59 universities in North America receiving the highest levels of funding from public biomedical sources similarly showed just 12 (20 per cent) scoring better than a C+ grade (Universities Allied for Essential Medicines UK, 2014).

There appears, at the moment, to be little correlation between global *need* and global *spend* in terms of health R&D. But what about within the 1 per cent of global spend that is actually allocated to neglected diseases? Is that distributed roughly according to need? A quick review of the figures in Table 8.3 would suggest, once more, the answer is 'no'. When the global burden of different 'neglected' diseases, measured in disability-adjusted life years (DALYs),[5] is divided by the total annual investment in R&D in those diseases, the resulting ratio varies from as little as 10 cents to as much as $1,400 per DALY.

Admittedly, the 'need' for health R&D does not solely depend on the relative scale of the global burden of disease. It is also necessary to understand the need for new knowledge and/or products for a given disease and what R&D has already been undertaken on it or is underway. Nevertheless, it can be expected that R&D needs would be broadly aligned to the scale of the health problem and that there should, therefore, be some rough correlation between R&D spend and the relative global burden of each particular disease (Viergever, 2013). As Table 8.3 demonstrates, even within the field of neglected diseases, R&D of new treatments does not necessarily conform to this expectation.

Table 8.3 Distribution of global health R&D funding across neglected diseases, 2011

Disease	*Global R&D funding ($ m)*	*Global burden of disease (million DALYs)*	*Funding per DALY ($)*
HIV	1,117	81.5	13.70
Malaria	596	92.7	7.20
Tuberculosis	584	49.4	11.80
Dengue	249	0.8	301.70
Diarrhoeal disease	169	89.5	2.00
Kinetoplastids	142	4.4	32.00
Bacterial pneumonia and meningitis	107	68.0–104.9	1.0–1.60
Helminths (worms and flukes)	90	12.3	7.30
Salmonella infections	48	17.1	2.80
Trachoma	10	0.3	31.10
Leprosy	8	0.006	1,400.90
Buruli ulcer	6	NA	–
Rheumatic fever	1	10.1	0.10

Source: Viergever, 2013: 2

One reason why there is so little relationship between R&D need and R&D spend, particularly within the area of neglected diseases, is that there is no functional global mechanism to either prioritize health R&D investments or to coordinate and ensure there is no duplication of effort. The World Health Organization has admitted as much, saying:

> One weakness of the current global health R&D efforts is the absence of quality information that provides a comprehensive overview on what is being supported, who is supporting it, how it is being supported and where it is being supported. Also lacking is the knowledge and capacity to set priorities at a high level and the extent to which many countries can collect and analyse this data in order to manage their own health research systems. (WHO, 2013)

There have been attempts to address this. In 2012 the WHO Consultative Expert Working Group on Research and Development called for three actions to fix the problem (WHO, 2012b):

1. All countries should commit to spend at least 0.01 per cent of GDP on government-funded R&D devoted to meeting the health needs of developing countries.
2. A global health R&D observatory should be established to collect and analyse data, including material on financial flows to R&D, the R&D pipeline, and lesson learning.

3. An internationally binding convention should be established that would provide a coordinating mechanism for identifying global health R&D priorities from 2) above and allocating funding from commitments under 1) above.

Agreement was reached at a World Health Assembly in November 2012 to develop norms and standards for the classification of health R&D, to establish the proposed Global Health R&D Observatory, and to implement a few health R&D demonstration projects. But US and European negotiators, facing financial constraints at home, stalled discussion on a binding convention and on any agreement around pooled funding, meaning the mechanism lacked the resources to make substantive progress on directing global health R&D. Discussions around a pooled funding mechanism that would provide an effective, coordinated, and sustainable source of funding for identified global health R&D priorities were postponed by member states until at least 2016 (Love, 2012), and commentators are sceptical about whether such funding will eventually be made available.

In addition, there have been a range of attempts to find ways and means of correcting the massive market failure to align private investment in the production of new drugs with actual need. One approach often referred to is the advanced market commitment (AMC), which aims to accelerate commercial R&D for a product that would deliver a specific health impact (for example, a vaccine) by offering an enhanced price to suppliers and a guaranteed initial size of demand if they can supply that product to market. The first AMC, for a pneumococcal vaccine, is being overseen by the vaccine alliance GAVI which offers an enhanced price of $7 per dose for 20 per cent of supplies in return for producers agreeing to supply the vaccine in the long term at a maximum price of $3.50 per dose (WHO, 2012b). Launched in 2009, it attracted commitments totalling $1.5 bn from the governments of Italy, the UK, Canada, Russia, and Norway, along with the Bill and Melinda Gates Foundation. GlaxoSmithKline and Pfizer both produced vaccines at the $3.50 per dose cost threshold under the AMC, and the vaccine has since been rolled out in 25 countries.

The AMC approach holds some promise although its performance and limited application to date means there are issues to resolve. Questions have been raised, for example, as to whether the price agreed under this AMC was too high (Glassman, 2013), considering manufacturers already produced the vaccines for developed-country markets and therefore needed limited further R&D to refine the vaccines for these markets by, for example, adapting them to local strains and incorporating the heat tolerance necessary for tropical climates. Indeed, the relative simplicity of the task was reflected by the very short time it took between the AMC being agreed (June 2009) and the first vaccines to be distributed in Latin America (December 2010). An evaluation of the initiative four years after its commencement did conclude that a multinational company with costs at the low–medium efficiency end of the range could have probably developed the vaccine at the $3.50 price point and made an acceptable internal rate of return without the AMC support (Dalberg, 2013).

It should also be noted that the cost of the AMC doesn't represent the entire cost of distributing the drug to those who need it. While donors have funded the supplement payable to manufacturers, GAVI itself finances the actual purchase of the vaccines. The AMC adds $1.5 bn to GAVI's income, yet GAVI estimates that between 2010 and 2030 it will, in addition, have to devote more than five times that amount ($8.1 bn) to subsidizing country purchases. This can happen only if countries also spend $6.2 bn of their own resources on vaccine purchase. Thus the headline cost of the AMC is a fraction of the total cost of getting the drug to those who need it (WHO, 2012b).

Finally, no other AMCs for drugs or vaccines have been tried to date and it is thought that the cost of a commitment for a drug in a much earlier stage of development than the pneumococcal vaccine was would be considerably higher than the $1.5 bn required on this occasion.

Straightforward grant financing is another approach to creating incentives. The Gates Foundation funds a set of Global Grand Challenges, some of which are aimed at stimulating innovation in the health sector. Current calls under the programme include developing new, more sensitive malaria diagnostics, new vaccination approaches, and methods to allow women to self-screen for cervical cancer. Funding of up to $100,000 is available to successful applicants to help them move to a proof-of-concept stage (Gates Foundation, 2015). At the time of writing there is also a Gates Grand Challenge with no financial ceiling that aims to encourage innovation to 'develop manufacturing platforms that can transform vaccine production economics and produce vaccines at a final finished goods production cost of ≤ $0.15 per dose' (Gates Foundation, 2015: 2). To ensure the greatest impact on reducing inequality in the application of discoveries, the vaccines have to come from a specified priority list that targets vaccines for diseases of great global burden which are among the most costly to produce with current technologies.

Grant financing is a 'push' incentive, where the financer risks funding failures as well as successes. An AMC is a 'pull' incentive in that the financer only funds successful outcomes (in the case described, vaccination doses delivered at an agreed price) and the risk of failure stays with the company concerned. Another approach to a 'pull' incentive is the 'prize' approach. The Ansari X Prize for Suborbital Flight is one of the most famous examples of this in recent times. The Ansari X Prize offered $10 m to the first successful team to launch a suborbital spacecraft and recover and relaunch it within two weeks. A case study of the prize reported that competitors collectively spent more than $100 m in pursuit of a $10 m award (McKinsey, 2009). The current Longitude Prize competition, started in 2014, is another example, more specifically relevant to the health sector. The competition offers a £10 m prize that will reward the first competitor who can 'develop a point-of-care diagnostic test that will conserve antibiotics for future generations and revolutionise the delivery of global healthcare' (Longitude Prize, 2014). A review of prizes worth $100,000 or more in 2007 revealed that there had been significant growth in their number (315 in 2007 compared to 74 in 1997), but that they were focused mostly on aviation and space, science and engineering, and climate change, with few awards for clinical research. McKinsey believes that 'recent prize proposals – $80 bn for new drugs, $300 m for car batteries, $100 m for hydrogen energy – represent an important uptick in interest' (McKinsey, 2009), but to date no prizes large enough to stimulate the sort of investment required to create a new drug have come into being.

Clearly, with just 1 per cent of all health R&D funding going into research on the diseases that impact most on the populations of the developing world, there remains a desperate need for a global coordination mechanism and pooled funding along the line of that proposed by the WHO Consultative Expert Working Group on Research and Development, probably alongside a significant scaling up of AMCs, prizes, and other approaches to stimulating innovation where there is market failure.

Agriculture: funding technical innovation for low external-input farming

The great agricultural challenge facing the world today is how to feed the 1 billion people who are malnourished *and* prepare to feed a global population set to grow by 3 billion by the middle of the century, while coping with increasing shortages of land and water, and the likely impacts of climate change. Ideally, all this will be achieved while simultaneously using agriculture to fight poverty and improve the livelihood opportunities for millions of smallholder farmers.

Raising agricultural productivity is clearly essential, but two major transformations have to be achieved to do this. Around half the world's food is currently produced by smallholder farmers, and small to medium-sized family farms account for around three-quarters of total production in the developing world. Productivity on smallholder farms in the developing world is constrained by the poor-quality marginal lands they often occupy and the inability of farmers to afford external inputs. As a result, the productivity gap between farmers in the North and South has increased since the start of the Green Revolution in the 1960s. Table 8.4 illustrates the scale of some of these gaps, particularly in sub-Saharan Africa where the Green Revolution failed to deliver widespread change. Clearly, smallholder farming in the developing world needs to go through some sort of intensification process to shift yields closer towards those in the developed world.[6]

But conventional modern agriculture is facing a crisis, too, as was suggested in Chapter 4. Professor Pablo Tittonell, in his 2013 inaugural address on taking up the position of Chair in Farming Systems Ecology at Wageningen University, noted that the doubling of average yields of major food crops achieved over the past 50 years had involved increasing the total amount of external nitrogen brought in through fertilizer use by a factor of seven, trebling phosphorus applications, and doubling the amount of water used for irrigation (Tittonell, 2013). The relative cost of Green Revolution technologies, in terms of resource use, looks even higher when energy inputs are accounted for. As Tittonell notes, energy inputs into food production have increased 50 times compared to traditional agriculture.

Table 8.4 Examples of yield gaps for key crops in developing-world regions

Region/country	*Yield gap (relative to the rest of the world average, %)*					
	Maize	*Sorghum*	*Cassava*	*Rice*	*Wheat*	*Millet*
Eastern Africa	320	40	46	107	77	–
Central Africa	60	495	97	340	78	75
Southern Africa[1]	34	–	–	85	4	–
Western Africa	200	77	17	148	78	–9[2]
India	–	52	–	39	12	1

1. excluding South Africa
2. minus sign indicates local yields are higher than the rest of the world average.
Source: Elliot, 2010

> Feeding an average person in the developed world costs about 1500 litres of oil equivalents per year. More than 30 per cent of this energy is used in the manufacture of chemical fertilisers, 19 per cent for the operation of field machinery and 16 per cent for transport. ... To feed 9 billon people in 2050 with the current production means of conventional agriculture we will need ... 113,000 million barrels of oil per year, close to 8 per cent of the total world reserve ... In other words, producing food for 9 billion people with conventional agriculture will exhaust our global oil reserves in about 12 years. (Tittonell, 2013: 5)

Combine this with concerns about land degradation, large-scale ecological damage from inorganic pesticide use, and water-resource pollution from nitrogen and phosphorus runoff, and it seems highly unlikely that it will be viable to increase global agricultural productivity to the levels required by 2050 using current conventional agricultural practice based on technologies derived from the Green Revolution.

Figure 8.2 illustrates the two great transformations required in global agriculture if we are to feed the world in the future. A process of intensification of smallholder agriculture has to take place to significantly raise productivity levels and move poor farmers out of the poverty trap of low access to external inputs translating into low returns on their investment of labour. But that has to be achieved in a much more efficient manner and using far fewer external resources than has been the case with conventional agriculture post-Green Revolution. Meanwhile, conventional agriculture has to be 'detoxified' or 'ecologized' significantly to reduce dependency on fossil fuels and fossil fuel-based inputs and their associated pollution, while maintaining yields. In other words, the production curve in Figure 8.2 needs to be moved substantially to the left.

Clearly, there is a massive need for agricultural R&D to underpin these two transformations. A summary of what those R&D needs might be, based on the conclusions of the International Assessment of Agricultural Knowledge, Science and Technology for Development (IAASTD),[7] is given in Table 8.5. The

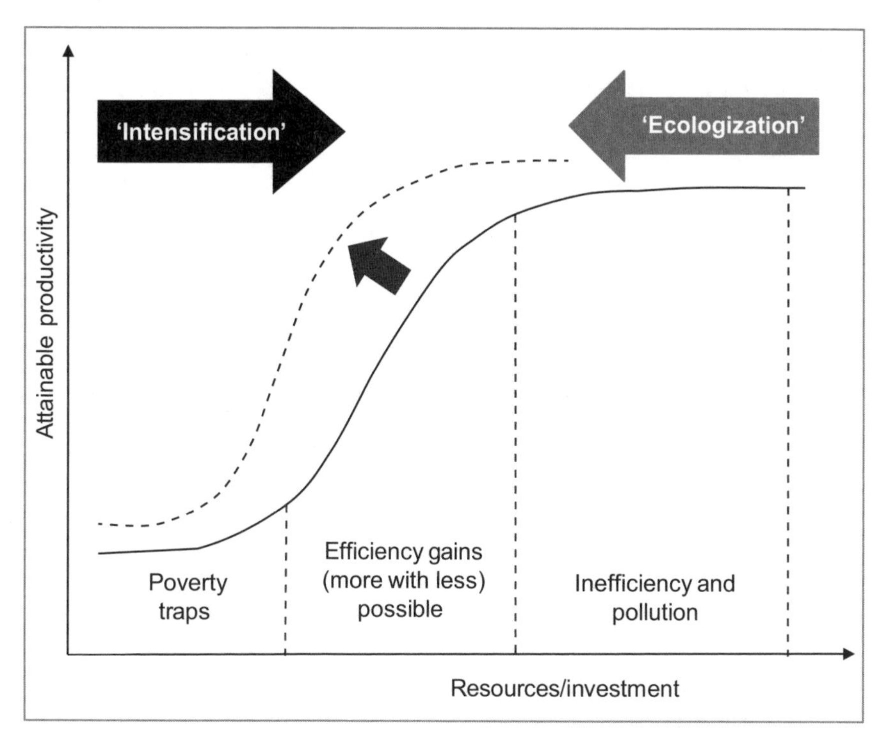

Figure 8.2 Attainable agricultural production per unit of land or person as a function of resource investment (capital and labour)

Source: Tittonell, 2013: 12

question is: does the focus of the current global programme of agricultural R&D match this need?

Unfortunately, the answer to this question has to be 'no'. As in the health sector, global R&D investments in the agriculture sector do not reflect need. In particular, as far as investment in the developing world goes, many of the trends seen in the health sector are repeated. Firstly, the private sector invests very little in research for the developing world. Agriculture presents particular challenges for inventors trying to profit from the benefits of their efforts. For crops that are self-pollinating, for example, farmers can reuse seed from year to year, making it difficult for seed companies to enforce patents and recover their costs. In this sector, as in the health sector, private R&D investment tends therefore to focus on areas where the benefits are more easily monetized: fertilizers, herbicides, insecticides, machinery, and hybrid seeds that have to be replaced every year or two. For instance, around a third of private investment in agricultural R&D globally is made in developing agricultural chemicals, most notably pesticides. Where investment in plant biology and breeding happens it tends to be on cash crops for export: palm oil, rubber, tea, vegetables, horticulture, and hybrid varieties of rice

Table 8.5 Agricultural R&D investment options outlined in the IAASTD Global Report

Agricultural knowledge system goal	*R&D investment required to:*	*Examples of R&D topics*
Environmental sustainability	1. Reduce the ecological impact of farming systems	Management practices; reduced use of fossil fuel, pesticides, fertilizer; biological substitutes for fossil fuels and chemicals
	2. Enhance systems that are known to be sustainable	Social science research on policies and institutions
	3. Support traditional knowledge	Non-conventional crops and breeds; traditional management systems
Hunger and poverty reduction	1. Target institutional change in organizations	Planning with a pro-poor perspective
	2. Include equity in planning and pro-poor policies	Access to resources; sharing of benefits from environmental services
Improving nutrition and human health	1. Improve nutritional quality and safety of food	Coexistence of obesity and micronutrient deficiency; pesticide residues; sanitary and phytosanitary measures
	2. Control environmental externalities	Pollution; overuse of antibiotics, pesticides; on-farm diversification
	3. Ensure better diagnostic data and response to epidemic disease	Increasing zoonotic diseases and dangers of pandemics; prediction of disease and pest migration with climate change
Economically sustainable development	1. Enhance research on water use and control of pests and diseases	Both affected by population growth and climate change
	2. Carry out productivity-enhancing research to save land and water as limiting factors	Total farm productivity benefits from higher yields per hectare and more crop per drop; need to address the most limiting factors
	3. Establish prices and incentives that promote proper social use of resources	Pricing policies and payment for ecosystem services to make land and water use more efficient
	4. Advance basic research in genomics, proteomics, and nanotechnology	Historically high rates of return on basic research; applications may spill over to developing countries in the future

Source: Beintema and Elliott, 2011: Table 9.3

and maize in Asia, for example, or in Africa, where private investment has been much lower, in cacao, tea, and coffee in Kenya (Naseem et al., 2010). Private investment in R&D, in practice, fails to focus on crops that are essential to livelihoods and food security in developing countries, such as sorghum, barley, or millet (because they are not essentially cash crops) or on open-pollinated maize varieties or vegetatively propagated crops, such as yams, cassava, potatoes, and sweet potatoes (because farmers control their reproduction). In low-income countries, excluding the largest nations like China and India, markets have the additional problem of often being too small or too poor to be attractive for private sector R&D investment, even in the case of cash crops.

Combined, these factors explain why private-sector investment accounted for 55 per cent of all investment in agricultural research in the US in 2000 but only 2 per cent of the agricultural R&D in developing countries. This means agricultural R&D in the developing world has to rely almost entirely on public funding, leading to a gross discrepancy in the resources available, as illustrated in Figure 8.3. The top 22 high-income countries spent a total of $22 bn on agricultural research in 2000, while sub-Saharan Africa (44 countries) spent just $1.5 bn and the entire developing world (117 countries) just $13.7 bn (Pardey et al., 2006).

In some parts of the developing world, investment in agricultural R&D is stagnant or even in decline. So while India and China managed annual growth rates of 4.4 per cent and 5.8 per cent respectively in their agricultural R&D spend over the 20 years to 2000, sub-Saharan Africa as a whole managed an annual growth in spend of only 0.6 per cent and half of those countries actually spent less in 2000 than they did in 1991 (Beintema and Elliott, 2011).

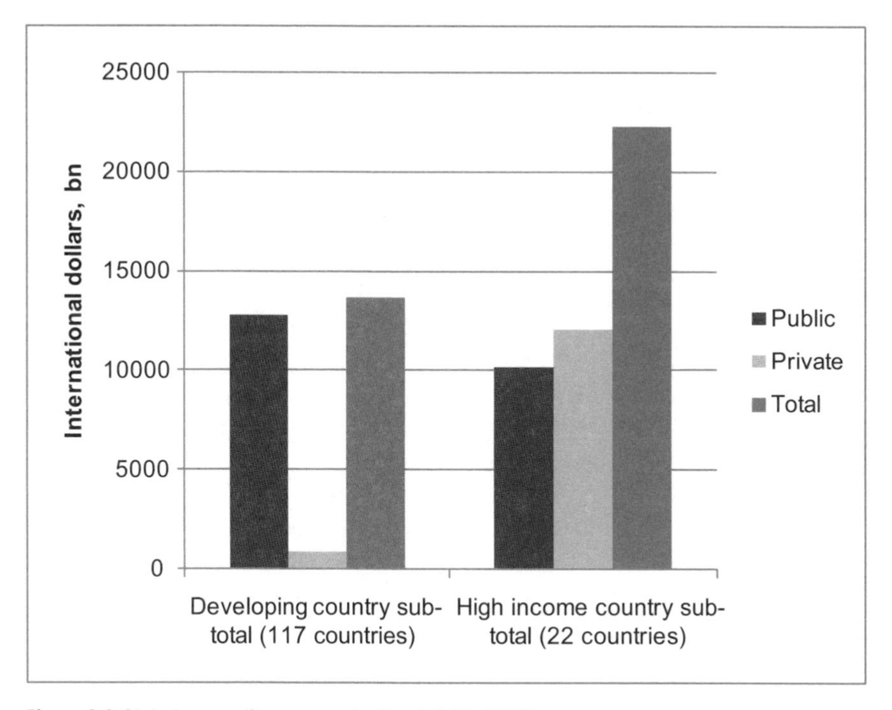

Figure 8.3 Global expenditure on agricultural R&D, 2000

Source: Pardey et al., 2006

Although there were some signs of improvement in the following decade, when public spend in agricultural R&D in sub-Saharan Africa increased by one a third in real terms, the growth was driven by a handful of countries, with 50 per cent accounted for by Nigeria and Uganda alone. So in 2011 only four countries in sub-Saharan Africa reported agricultural R&D spends of over $100 m a year, while roughly half the countries in the region reported annual investment levels below the $10 m mark (Beintema and Stads, 2014).

Meanwhile, in the developed world, the dominance of the market in shaping the focus of research on products where intellectual property rights can be enforced is itself a barrier to research into more sustainable forms of agriculture, such as agroecology, where the emphasis is on knowledge and techniques that are difficult to patent. It is hard to find solid evidence of the levels of investment in R&D into more sustainable forms of agriculture, but one estimate suggests 'that the research investment gap between organic and conventional agriculture since the onset of the green revolution, considering both public and private sector investments should be around 90 to 95 per cent' (Tittonell, 2013: 13).

Recipes vary for correcting the imbalance between actual and needed investment in global agricultural R&D. Proposals to increase private-sector investment into pro-poor agricultural R&D in the developing world concentrate on a similar range of mechanisms to those proposed for health R&D, including push mechanisms such as public grants to support private R&D, tax breaks, and public–private partnerships, and pull mechanisms such as prizes and advanced market commitments (see, for example, Naseem et al., 2010; Elliot, 2010), although there seem to be fewer examples of these mechanisms actually being used in developing-country contexts to date. On the public funding side, various proposals have been made including one for a minimum 5 per cent annual growth in agricultural R&D spending in low and middle-income countries over the next 10 years and an allocation of at least 1 per cent of agricultural GDP to public agriculture R&D, which would equate to a doubling of spend in sub-Saharan Africa (Beintema and Stads, 2014). Others note the history of positive impacts of the largest publicly funded global research partnership, the Consultative Group for International Agricultural Research (World Bank, 2010), and, in particular, the important role it plays as a funder of research in sub-Saharan Africa and its potential as a vehicle for change (Beintema et al., 2012). There is, however, a general lack of proposals in the literature for drivers that would specifically result in a shift in R&D investment towards agroecological approaches to food production, especially in the developed world. At the moment, it seems that we are a long way from the path to a sustainable and equitable food production system.

Driving sustainable pro-poor innovation: the need for change

The discussion concerning the health and agriculture sectors illustrates just two examples of the way global drivers of technology R&D do not respond to societal challenges or focus on improving the lives of those least well off in society. As such, they stand as examples of how our global systems of technology innovation today are unjust. A similar analysis could have been carried out in other sectors. In the energy-generation sector, for example, the disproportionally high $550 bn global annual public subsidy to support fossil fuels exploration and production dwarfs the $120 bn of equivalent subsidy for renewable energy technology development, ensuring the impetus remains focused on the technologies most likely to lead to a breach of planetary boundaries rather than on those that would help us move to a more sustainable future.

It is clear that a wholesale shift in the way investment is allocated to technology R&D is required, both in the public and private sectors, if we are going to see technology address the challenges of achieving a universal social foundation while remaining within ecological planetary boundaries. This needs to include but also extend beyond risk management through use of the precautionary principle to a new set of governance processes that manage the direction of the global technological innovative effort. This will be returned to at the end of Part II of this book, but before that the next two chapters focus on two issues that have been touched upon but which merit further discussion in their own right: the role of intellectual property rights in facilitating or restricting access to and innovation of technologies; and the role of the state in technology innovation.

Notes

1. Geoengineering refers to the deliberate large-scale manipulation of environmental processes in an attempt to reduce the impacts of global warming. These can include proposed techniques such as the blocking of sunlight with space-based mirrors or the seeding of the oceans to increase CO_2 absorption.
2. For example, the article talks about adversaries in France where the Academy of Sciences recommended 'that the precautionary principle should not be attached or included in constitutional texts or in the high level "organic laws", to the extent that doing so could induce a deleterious effect with disastrous consequences for future progress of well-being, health and our environment' (cited in ParisTech Review, 2014).
3. See Universities Allied for Essential Medicines UK (2012) for a description of the full methodology.
4. The evaluation defines neglected diseases as a group of infections that together affect over 1 billion people in the world and are estimated to account for at least 10 per cent of the global disease burden. As they are diseases that are almost exclusively found among the world's poorest populations, there is little incentive for profitable R&D investment to create new treatments, vaccines, or diagnostics for them. The list of diseases itself is based on the 2011 version of the G-FINDER Report (Policy Cures, 2011: 14), an annual analysis of global funding availability for research into neglected diseases

carried out by Policy Cures and sponsored by the Bill and Melinda Gates Foundation. The full list is provided in Appendix 2.

5. Disability-adjusted life years is a formulation the WHO uses to express the overall disease burden, calculating the number of years of life lost due to death or restrictions resulting from illness compared to someone of perfect health.
6. The supposed 'yield gaps' shown here are based on maximum achievable yields using Green Revolution technologies. This book consistently argues for an agroecological approach, which is unlikely to achieve comparable yields on individual crops to Green Revolution approaches (but might achieve better productivity in terms of total food produced per hectare, given multicropping approaches). The argument made throughout this book is thus not to close these yield gaps entirely, but to significantly reduce them by improving the productivity of smallholders using low-input agroecological approaches.
7. The IAASTD report, published in 2009, was the result of a five-year review to assess the impacts of past, present, and future agricultural knowledge, science, and technology on: the reduction of hunger and poverty; improvement of rural livelihoods and human health; and equitable socially, environmentally, and economically sustainable development. It was the result of a global consultation initiated by the FAO and the World Bank which drew on the work of hundreds of experts from all regions of the world who participated both in the review and peer-review process of the report (McIntyre et al., 2009: vi).

References

Ahteensuu, M. (2007) 'Rationale for taking precautions: normative choices and commitments in the implementation of the precautionary principle', in *Social Contexts and Responses to Risk, Streams Programme: Risk and Rationalities Conference* [online], 29 March, <http://www.kent.ac.uk/scarr/events/ahteensuu.pdf> [accessed 18 September 2015].

Beintema, N. and Elliott, H. (2011) 'Setting meaningful investment targets in agricultural research and development: challenges, opportunities and fiscal realities', in P. Conforti (ed.), *Looking Ahead in World Food and Agriculture Perspectives to 2050*, pp. 347–87, Rome: Food and Agriculture Organization of the United Nations.

Beintema, N. and Stads, G.-J. (2014) 'Agricultural R&D: is Africa investing enough?', in A. Marble and H. Fritschel (eds), *2013 Global Food Policy Report*, pp. 53–62, Washington, DC: International Food Policy Research Institute.

Beintema, N., Stads, G., Fuglie, K., and Heisey, P. (2012) *ASTI Global Assessment of Agricultural R&D Spending: Developing Countries Accelerate Investment*, Washington, DC: IFPRI.

Commission on Health Research for Development (1990) *Health Research – Essential Link to Equity in Development*, Oxford: Oxford University Press.

Dalberg (2013) *The Advance Market Commitment for Pneumococcal Vaccines: Process and Design Evaluation*, Geneva: GAVI Alliance.

The Economist (2015) 'Clinical trial publishing game, from The Economist', 6 October, *+AllTrials*, <http://www.alltrials.net/news/the-economist-publication-bias/> [accessed 3 November 2015].

Elliot, K. (2010) 'Geoengineering and the precautionary principle', *International Journal of Applied Philosophy*, 24: 237–53 <http://dx.doi.org/10.5840/ijap201024221>.

European Environment Agency (2001) *Late Lessons from Early Warnings: The Precautionary Principle 1896–2000*, Luxembourg: European Environment Agency.

European Environment Agency (2013) *Late Lessons from Early Warnings: Science, Precaution, Innovation*, Copenhagen: European Environment Agency.

Gates Foundation (2015) 'Grand challenge: innovations in vaccine manufacturing for global markets', *Bill & Melinda Gates Foundation*, <http://gcgh.grandchallenges.org/sites/default/files/grant-opportunity-rfp/Vaccine%20Manufacturing%20Grand%20Challenge%20%281%29.pdf> [accessed 6 November 2015].

GBD 2013 Risk Factors Collaborators (2015) 'Global, regional, and national comparative risk assessment of 79 behavioural, environmental and occupational, and metabolic risks or clusters of risks in 188 countries, 1990–2013: a systematic analysis for the Global Burden of Disease Study 2013. *The Lancet*, 386: 2287–323 <http://dx.doi.org/10.1016/S0140-6736(15)00128-2>.

Gee, D. (2013) 'More or less precaution?', in *Late Lessons from Early Warnings: Science, Precaution, Innovation*, pp. 643–69, Luxembourg: European Environment Agency.

Glassman, A. (2013) 'Is the price right? Evaluating advanced market commitments for vaccines' [blog], 14 June, Center for Global Development, <http://www.cgdev.org/blog/price-right-evaluating-advanced-market-commitments-vaccines> [accessed 6 November 2015].

Hansen, S. and Tickner, J. (2013) 'The precautionary principle and false alarms – lessons learned', in *Late Lessons from Early Warnings: Science, Precaution, Innovation*, pp. 17–45, Copenhagen: European Environment Agency.

Karinen, R. and Guston, D.H. (2010) 'Towards anticipatory governance: the experience with nanotechnology', in M. Kaiser, M. Kurath, S. Maasen, and C. Rehmann-Sutter (eds), *Governing Future Technologies: Nanotechnology and the Rise of an Assessment Regime*, pp. 217–32, Sociology of the Sciences Yearbook 27, Dordrecht: Springer Science+Business Media.

Lofstedt, R. (2004) 'The swing of the regulatory pendulum in Europe: from precautionary principle to (regulatory) impact analysis', *Journal of Risk and Uncertainty*, 28: 237–60 <http://dx.doi.org/10.1023/B:RISK.0000026097.72268.8d>.

Longitude Prize (2014) 'Antibiotics: how can we prevent the rise of resistance to antibiotics?', *Longitude Prize*, <https://longitudeprize.org/challenge/antibiotics> [accessed 22 January 2016].

Love, J. (2012) 'WHO negotiators propose putting off R&D treaty until 2016', 28 November [blog], *Knowledge Ecology International*, <http://keionline.org/node/1612> [accessed 5 November 2015].

Mandel, G.N. and Gathii, J.T. (2006) 'Cost-benefit analysis versus the precautionary principle: beyond Cass Sunstein's laws of fear', *University of Illinois Law Review*, 1037–80 <http://ssrn.com/abstract=822186>.

Mazzucato, M. (2013) *The Entrepreneural State: Debunking Public vs. Private Sector Myths,* London: Anthem Press.

McIntyre, B.D., Herren, H.R., Wakhungu, J., and Watson, R.T. (2009) *Agriculture at a Crossroads: Synthesis Report,* Washington, DC: International Assessment of Agricultural Knowledge, Science and Technology for Development.

McKinsey (2009) *'And the Winner Is ...' Capturing the Promise of Philanthropic Prizes,* Sydney: McKinsey & Company.

Mellon, M. and Rissler, J. (2003) 'Environmental effects of genetically modified food crops: recent experiences', presented at *Genetically Modified Foods – the American Experience,* Copenhagen: Royal Veterinary and Agricultural University, 12–13 June.

Naseem, A., Spielman, D.J., and Omano, S.W. (2010) 'Private-sector investment in R&D: a review of policy options to promote its growth in developing-country agriculture', *Agribusiness,* 26: 143–73 <http://dx.doi.org/10.1002/agr.20221>.

Pardey, P.G., Beintema, N.M., Dehmer, S., and Wood, S. (2006) 'Agricultural research – a growing divide?', Washington, DC: IFPRI.

ParisTech Review (2014) 'Is it really possible to enforce the precautionary principle?', 28 March, *ParisTech Review,* <http://www.paristechreview.com/2014/03/28/enforce-precautionary-principle/> [accessed 18 September 2015].

Policy Cures (2011) *G-FINDER 2011. Neglected Disease Research and Development: Is Innovation Under Threat?* London: Policy Cures <http://policycures.org/downloads/g-finder_2011.pdf>.

Powell, R., Clarke, S., Sheehan, M., Douglas, T., Foddy, B., and Savulescu, J. (2010) 'The ethics of geoengineering', *Practical Ethics,* Oxford University, <http://www.practicalethics.ox.ac.uk/__data/assets/pdf_file/0013/21325/Ethics_of_Geoengineering_Working_Draft.pdf> [accessed 22 January 2016].

Røttingen, J. et al. (2013) 'Mapping of available health research and development data: what's there, what's missing, and what role is there for a global observatory?', *The Lancet,* 382: 1286–1307 <http://dx.doi.org/10.1016/S0140-6736(13)61046-6>.

Solon, O. (2013) 'Bill Gates: capitalism means male baldness research gets more funding than malaria', 14 March, *Wired,* <http://www.wired.co.uk/news/archive/2013-03/14/bill-gates-capitalism> [accessed 24 April 2015].

Tedsen, E. and Homann, G. (2013) 'Implementing the precautionary principle for climate engineering', *Carbon & Climate Law Review,* 2: 90–100.

Tittonell, P.A. (2013) 'Farming systems ecology: towards ecological intensification of world agriculture', 16 May, Wageningen: Wageningen University <https://www.wageningenur.nl/en/show/Towards-ecological-intensification-of-world-agriculture.htm> [accessed 17 March 2016].

Universities Allied for Essential Medicines UK (2012) 'UK Universities Global Health Research League table methodology', *Global Health Grades,* <http://globalhealthgrades.org.uk/cms/assets/uploads/2012/10/FullMethodologyPDF.pdf> [accessed 17 March 2016].

Universities Allied for Essential Medicines UK (2014) 'University Global Health Research League table', *Global Health Grades,* <http://globalhealthgrades.org.uk/about/> [accessed 3 November 2015].

Viergever, R.F. (2013) 'The mismatch between the health research and development (R&D) that is needed and the R&D that is undertaken: an overview of the problem, the causes, and solutions', *Global Health Action*, 6: 22450 <http://dx.doi.org/10.3402/gha.v6i0.22450>.

WHO (2012a) *Global Report for Research on Infectious Diseases of Poverty*. Geneva: World Health Organization.

WHO (2012b) *Report of the Consultative Expert Working Group on Research and Development: Financing and Coordination*. Geneva: WHO.

WHO (2013) *Draft Working Paper 2: Coordination and Priority Setting in R&D to Meet Health Needs in Developing Countries*. Geneva: WHO.

World Bank (2010) *Forty Findings on the Impact of CGIAR Research 1971–2011*, Washington, DC: World Bank.

CHAPTER 9

Intellectual property rights: part of the solution or part of the problem?

Why patent?

In the absence of regulation or legislation ruling otherwise, technology is mostly non-excludable. My use of a mobile phone or a plough does not exclude anyone else from also using a phone or plough. This can be a problem from the perspective of innovation, as the inventor has to invest time, effort, and money in the innovative effort, but may not be able to recoup their costs or make a profit once an invention is in the public domain and anyone can copy it. It was for this reason that intellectual property rights (IPR) or patents were introduced. Patents generally confer on the inventor the exclusive right either to use their technology or to license others to use it. This monopoly is generally granted for a fixed period of time, often up to 20 years.

Patents are territorial and are issued by national patent offices. The World Intellectual Property Organization (WIPO) defines the conditions that should be met in order for a patent to be granted as (WIPO, 2015):

- The invention must show an element of novelty; that is, some new characteristic which is not known in the body of existing knowledge in its technical field ...
- The invention must involve an 'inventive step' or 'non-obviousness', which means that it could not be obviously deduced by a person having ordinary skill in the relevant technical field.
- The invention must be capable of industrial application, meaning that it must be capable of being used for an industrial or business purpose beyond a mere theoretical phenomenon, or be useful.
- Its subject matter must be accepted as 'patentable' under law. In many countries, scientific theories, aesthetic creations, mathematical methods, plant or animal varieties, discoveries of natural substances, commercial methods, methods for medical treatment (as opposed to medical products), or computer programs are generally not patentable.
- The invention must be disclosed in an application in a manner sufficiently clear and complete to enable it to be replicated by a person with an ordinary level of skill in the relevant technical field.

In theory, patents offer two important means to stimulate technological innovation. Firstly, they provide protection for the original inventor and encourage investment in innovation because that protection offers the possibility of achieving a financial return through a monopoly. Secondly,

http://dx.doi.org/10.3362/9781780449043.010

patents require public disclosure of the details of the invention which, otherwise, may not be in the public domain (keeping secret the details of how an invention works is another way of protecting its value to the inventor). Disclosure means that others can build on that knowledge and perhaps develop further inventions and innovations themselves.

Do patents encourage innovation?

Patenting activity is seen as so central to technology innovation today that it is often used as a proxy indicator of levels of innovation in an economy or a company. The UK government introduced a policy in 2013, for example, to reduce the rate of corporate tax on income derived from patents, on the assumption that this would stimulate more innovation which would, in turn, be good for the economy. Likewise, venture capitalists are known to use the number and frequency of patents registered as an indicator for which companies to invest in (Mazzucato, 2013: 51). But there is a growing school of thought that questions the link between patents and innovation and raises the possibility that the IPR edifice may actually be an impediment to the development of new and useful technologies. There are several reasons why these doubts are being raised.

Firstly, there is a lot of evidence that suggests numbers of patent registrations are *not* good proxies of levels of innovation. For a start, in an environment where venture capitalists use patents as a way of identifying suitable investment targets, there is a perverse incentive for companies to register as many patents as possible to attract investment, whether or not this leads to any real productive activity. More generally though, there may be no relationship between patent volumes and productivity. A paper by Boldrin and Levine arguing the case against patents shows that the number of patents issued in the United States rose from 59,715 in 1983 to 244,341 in 2010. Despite this more than quadrupling of the rate of patent registration in less than 30 years, the researchers show that 'neither innovation nor research and development expenditure nor factor productivity have exhibited any particular upward trend' during that period, leading them to conclude that 'there is no empirical evidence that ... [patents] ... serve to increase innovation and productivity, unless productivity is identified with the number of patents awarded – which, as evidence shows, has no correlation with measured productivity' (Boldrin and Levine, 2013).

Secondly, there is evidence of parasitical behaviour that adds very little real value to society around patents. The existence of 'patent trolls' is one such example. These are companies that exist solely to buy up patents and extort rent from others deemed (often on flimsy grounds) to be using similar technology, by threatening lawsuits for infringement of their patents. The development of 'patent thickets' is another example. This occurs when a company registers a wide range of speculative patents in an effort to prevent competition from being able to work on a particular problem. Indeed, some of this parasitical behaviour may actually incur significant costs to society. US official statistics are said to show that patent trolls were responsible for up to 62 per cent of all infringement lawsuits from 2011 to 2013, and other research estimates that trolls were responsible for around half a trillion dollars of lost wealth between 1990 and 2010 (Kenny and Barder, 2015).

Thirdly, there is evidence that patents not only fail to drive innovation in weak markets, but also significantly under-represent what constitutes the bulk of the global innovative effort. In essence, when the combined gross national income of low-income countries ($634 bn) is just over 1 per cent of that of high-income countries ($51 tn), it is almost inevitable that technology R&D will focus largely on the problems of the 'rich' world, where markets provide strong demand and a patent will deliver a good return on investment. In addition, the focus on patents as the measure of innovation ignores the fact that in weak markets technical progress may be made more by the adoption of innovation from elsewhere than by local invention. In *Science and Technology for Development*, James Smith comments on this aspect of the value of IPR for less developed countries, noting: 'There are relatively few benefits in terms of stimulating local innovation in developing countries, as technological activity in such countries tends to focus on learning to use imported technologies rather than to innovate new technologies' (Smith, 2009: 89). He goes on to say, 'evidence suggests that strong IPR only begins to benefit countries with per capita incomes above $7,750, as they move away from building local capabilities through copying and begin to engage in more innovative activities' (Smith, 2009: 89). Indeed, other authors argue that this is not only the case for weak markets but that patent activity in general underestimates, and so is a poor reflection of, overall levels of technological development activity in an economy because

> patents refer to inventions, not innovations, and are used much more intensively in some industries than others. In fact, the global novelty requirement associated with patents implies that minor innovations/adaptations, which arguably make up the bulk of innovative activity world-wide, will not be counted since these are simply not patentable ... (Fagerberg et al., 2009: 27)

TRIPs, patents, and the negative impacts on developing countries

Perhaps the biggest concern with IPR is not the failure of patents to drive constructive innovative behaviour, or even their failure to account for the bulk of innovation, but the potential destructive impact they can have on innovation efforts. The economists Joseph Stiglitz and Claude Henry, in a 2010 paper on intellectual property and sustainable development, attribute this destructive force to the way patent legislation developed in the US during the 1980s (Henry and Stiglitz, 2010). In 1982 the US Congress created the Court of Appeals for the Federal Circuit, which specialized in IP matters. This was done, according to Stiglitz and Henry, as a response to growing pessimism about the competitiveness of the US economy and with a view to supporting 'an approach that would be systematically sympathetic to the defence and promotion of intellectual property' (Henry and Stiglitz, 2010: 242). The result was a dramatic increase in the number of rulings on patent infringements in favour of the patent holder. This, combined with a simultaneous starving of funds to the US Patent and Trademark Office, which unintentionally prevented officers there from properly examining patent applications, led to a steep rise in both the number of patent applications made and the number accepted.

The conditions for a patent (laid out at the beginning of this chapter) started to be compromised to the extent that, today,

> Patents are routinely granted to submissions devoid of any novelty or with insignificant original contributions ... [or] to parties that are not the real innovators. Overlapping patents are granted, which is a sure recipe for igniting inextricable conflicts, exacerbating the already oppressive problem of the patent thicket ... Patents that are broader than they should be are routinely granted. (Henry and Stiglitz, 2010: 242)

Patents also started to be issued to US holders for traditional knowledge or products such as basmati rice, neem oil, or the healing powers of turmeric, again breaching the spirit of the conditions for patents outlined by the WIPO.

Against this background, and driven largely by a 'small group of American lawyers and chief executives of large firms, active mostly in the entertainment industry ... software and life sciences' (Henry and Stiglitz, 2010: 243), pressure began to mount during the Uruguay Round of talks (1982–94) under the General Agreement on Tariffs and Trade (GATT) to find a mechanism to enforce IPR globally by incorporating them in a global trade agreement. A bargain was struck between developed and developing countries that was supposed to lead to greater access for developing-country goods to developed-economy markets in return for liberalization of finance markets and enforceable global rules on IPR. In the end, the West reneged on most of its side of the bargain, but did 'succeed in forcing the TRIPS (Trade Related Aspects of Intellectual Property Rights) on a reluctant developing world' (Henry and Stiglitz, 2010: 243).

The impact of TRIPS on developing countries' access to technology has been hinted at in earlier chapters. It is often referred to in the context of health and the impact it has had on the costs of health-care in developing countries, as it effectively extended the mandate of (mainly US and European) pharmaceutical companies' patents to developing countries, blocking the right to manufacture cheaper generic drugs. In a 2008 paper Stiglitz gives the example of HIV/AIDs drugs. He quotes the cost of a year's treatment using western brand-name drugs as around $10,000 and notes that this puts treatment way beyond affordability for a person with AIDS in a developing country, whereas generic medicines, which sell for less than $200, might be affordable. Stiglitz suggests, dramatically, that, 'When the trade ministers signed the TRIPS agreement in Marrakesh in the spring of 1994, they were in effect signing the death warrants on thousands of people in sub-Saharan Africa and elsewhere in the developing countries' (Stiglitz, 2008: 1701).

The Uruguay round of GATT talks did attempt to insert some safeguards into the TRIPS agreement, with rules permitting compulsory purchase of licences to allow for production of medicines where there are significant threats to a nation's health. There are some instances of this facility being used, for example, in the issuing of compulsory licences for the production of generic HIV/AIDS antiretrovirals in South Africa, Zimbabwe, and Kenya (Musungu and Oh, 2005). But aggressive pursuit of patent protection, particularly by the US, has led to a number of regional trade agreements that actually add further burdens of patent protection, beyond the TRIPS requirements, on developing countries. These agreements have become known as TRIPS+ and have attracted fierce criticism for the additional costs they potentially place on developing economies. A 2006 report on a US–Colombia trade deal, for instance, highlighted the fact that, as a result of the TRIPS+ nature of the agreement, the South American nation would need to spend an additional $919 bn by 2020 just to maintain the same level of medical care it had at present (Carter, 2012).

Asymmetries of power

Others have noted the asymmetry of IPR enforcement in the health arena. While developing countries have to pay more for drugs, the drug companies concerned invest little in the diseases that are more likely to affect the poor in developing countries. At the same time there is little protection given to indigenous IPR for the traditional knowledge of developing countries, as evidenced by the drug companies' 'opposition to paying for the value of the knowledge associated with the genetic material obtained from developing leading to the refusal of the United States and other advanced industrial countries to sign the Convention on Biological Diversity' (Henry and Stiglitz, 2010: 244).

In his book on intellectual property, biodiversity, and sustainable development, the economist Martin Khor discusses the misappropriation of traditional biodiversity knowledge or 'biopiracy', citing it as one of the most 'complex problems facing the future of traditional knowledge' (Khor, 2002). In most developing countries there has been no tradition of private ownership of knowledge concerning biodiversity, such as that related to agriculture, livestock, fishing, or the use of naturally occurring plants with medicinal properties. Knowledge concerning the cultivation of seeds or the use of plants or the breeding of animals has been shared between communities and individuals and has been one of the key factors in maintaining biodiversity in farming systems and in natural habitats. Khor argues that this system of community sharing and collaborative innovation is being challenged by IPR and the TRIPS regime, which together create a new system to exert private ownership rights over knowledge. As a result of the complexity and cost of the process of registering ownership and obtaining those rights, large corporations or institutions that have the necessary financial resources and legal expertise are favoured over local communities, who find it all but impossible to participate in the system or to obtain the rights they should be entitled to. Khor backs this assertion up with examples showing:

- attempts to create huge market monopolies through the registration of very broad patents which contain 'bio-piracy elements', one example being the US company Mycogen's European patent that covers the insertion of 'any insecticidal gene in any plant' and which is based on *Bacillus thuringiesis* (Bt), a naturally occurring soil bacterium which produces a protein fatal to many insects that consume it and has been used as a biological pesticide by farmers since the 1940s;
- attempts to patent traditional uses of medicinal plants, including a Japanese company patenting various traditional Filipino herbal remedies, American scientists patenting a protein from a native species of Thai bitter gourd after Thai scientists found that compounds from that variety could be useful against the AIDS virus, and a (failed) attempt by American scientists to patent the use of turmeric for healing wounds (in India a traditional remedy for sprains, inflammatory conditions, and wounds for centuries);
- patents held on gene sequences for staple crops, mostly by American and Japanese companies, including rice, maize, potato, and wheat varieties.

One of the key issues here is that if it is a seed that is patented, it could lead to a situation where farmers in developing countries, possibly including the country from which the seed material originated, are forced to 'buy and use, but not save and reuse' seed and thus incur greater costs. There may also be restrictions on the ability of countries to conduct further research using the seed.

Alternatives to the existing patent system

Intellectual property rights regimes, at least in their current form, are no longer fit for purpose. It is often not possible to see a direct link between the lodging of patents and the impetus for genuine technological innovation. Indeed, there is significant evidence to make the counter argument that patents are more often a blockage or a brake on the innovative effort with, particularly in the US, patent trolls and patent thickets meaning anyone actually trying to apply a new invention in practice faces the very real risk of incurring a lawsuit and significant punitive damages as a result of the parasitic behaviour of others. Moreover, the incorporation of IPR into GATT via TRIPS has led to huge asymmetries of power between the developed and developing world, to the great disadvantage of the latter. Even where safety features such as compulsory licensing have been incorporated in TRIPS, there is evidence that developing countries come under significant pressure not to use them and that bilateral and regional trade deals between developing and developed countries actually incorporate even stricter interpretation of IPR (TRIPS+). The scale of the asymmetry of power is illustrated in corporate attempts to 'privatize' centuries-old indigenous knowledge and naturally occurring substances that should form part of the resource base of poorer nations, a process often referred to in the literature as 'biopiracy'.

If the current regime is unfit for purpose, what is the alternative? Patents were introduced in response to a very real problem – the disincentive to invest in innovation when the innovator is unlikely to recoup the cost and make a profit from the innovation. Stiglitz, while heavily critical of patents, does not call for their abolition but for a downgrading of their relative importance and an increase in the use of alternative incentives, such as prizes and grants, as part of a portfolio of stimuli for an innovation system (Stiglitz, 2008). Stiglitz identifies five attributes or tasks associated with innovation:

1. *selecting* which research projects are to be done and who will be the researchers;
2. identifying how the research will be *financed*;
3. managing (and deciding who bears) the *risks* associated with research, which inherently is an activity with uncertain outcomes;
4. creating the *incentives* for individuals and firms to innovate;
5. ensuring the results of research are *disseminated* and *used*.

He goes on to use these attributes to compare the advantages and disadvantages of prizes and grants to patents (see Table 9.1). In short, Stiglitz sees patents as having the potential to offer high levels of incentive to innovate, although that is tempered by high levels of risk of litigation in today's environment. They also have the advantage of being completely self-selecting (the patentee decides what to innovate and bears the risk that the research will not produce returns), which means they can encourage innovation in all areas. A major disadvantage is that, by granting, in effect, a licence via the patent for the patentee to recover costs by charging a tax on the user of the technology, patents can be highly distortionary and inequitable (for example, a patent on a medicine allows it to be sold by the patent holder at a cost higher than the marginal cost of its production, thus, in essence, taxing someone who is ill for their illness). The utilization of knowledge is most efficient in an economy when it flows freely to all players, so the 'tax' levied by patent holders also acts as a brake on the efficient dissemination of knowledge.

Prizes have an advantage in that they can be constructed to ensure the winning technology is available to all as long as the prize is big enough to cover the real cost of developing the technology in the first place. They can also create incentives for innovation in socially or environmentally important areas that would not otherwise occur and do not necessarily impose a distortionary or inequitable subsequent tax on users of the resulting technology. They are partly self-selecting in that innovators choose whether

Table 9.1 Comparing alternative innovation incentive systems

Attribute	*Innovation system*		
	Patent	*Prize*	*Government-funded research*
Selection	Decentralized, self-selection	Decentralized, self-selection	Bureaucratic
	Lacks coordination	Lacks coordination	More coordination possible
Finance	Highly distortionary and inequitable	Can be less distortionary and more equitable	Most efficient
Risk	Litigation risk	Less risk	Least risk
Innovation incentive	Strong but distorted	Strong, less distorted	Strong non-monetary incentives
		Requires well-defined objectives	
Dissemination incentive	Limited – monopoly	Strong – competitive markets	Strong

Source: Stiglitz, 2008: 1722

to take part or not and only successful innovations are awarded a prize. One disadvantage of prizes is that, unlike patents, in selecting the area for which a prize is to be awarded, the prize-giver forgoes the opportunity to incentivize innovation in other unforeseen areas that may have delivered even greater social or environment return on investment. Another disadvantage is that, in some areas of research, notably biomedical research into new drugs, the costs of development can be enormous, meaning that the scale of prize required to act as an incentive may be untenable.

One interesting commercial use of prizes to stimulate innovation is in data analytics. Kaggle, for example, is an online company that crowd-sources predictive modelling solutions for companies via competitions. In its own words:

> Many organizations don't have access to the advanced machine learning that provides the maximum predictive power from their data. Meanwhile, data scientists and statisticians crave real-world data to develop their techniques. Kaggle offers companies a cost-effective way to harness this 'cognitive surplus' of the world's best data scientists. Our vibrant community comprises experts from many quantitative fields and industries (science, statistics, econometrics, math, physics). They come from over 100 countries and 200 universities. In addition to prize money and data, they use Kaggle to learn, network, and collaborate with experts from related fields. (Kaggle, 2015)

A company with a problem to solve prepares the data and a description of the problem. Data scientists participating in the competition then try out different techniques and compete against each other to produce the best models. Solutions submitted are scored immediately based on how closely their predictions match a hidden solution file and the scores are displayed on a live leader board, allowing teams or individuals to work further on a solution to improve it and resubmit. Once the deadline passes, the company hosting the competition pays the individual or team with the highest score the prize money in exchange for all IPR for the solution. At the time of writing, Kaggle was hosting six competitions to solve problems ranging from a model to predict sales for a European drug store chain (prize $35,000) to a model to predict property rental prices in Australia for Deloitte (prize $100,000).

The more traditional approach of awarding research grants puts all the risk with the grant-giver, who chooses both the topic and who will do the research, and so relies entirely on the grant-giver's ability to 'pick a winner'. It is generally an input-based contract which pays whether the research yields a useful product or not. For the innovator, it provides the least risk environment and the greatest opportunity for dissemination of results. For the innovation system as a whole, it allows the greatest amount of coordination and the ability to avoid duplication of effort and waste of resources between different researchers.

Patent pools are also cited by some as a better way of managing IPR. Patent pools are typically used where competing businesses all hold patents for different technologies that would need to be combined to produce a new core technology. This might result in a stalemate where a new technology could not advance but, by pooling patents and licensing each other to use them, a group of companies can collaboratively use their shared IPR to develop new products. Interoperability requirements in the software and electronics industry is one set of motivations for the development of patent pools, for example, the patent pool for MPEG that covers the patents necessary to work with the MPEG international standard for encoding and compressing video images (Bristows, 2009).

The concept of patent pools has been used with a social aim in mind. The United Nations agency UNITAID established a voluntary patent pool for HIV/AIDS therapies in 2010. UNITAID itself was established in 2006 to tackle inefficiencies in markets for drugs for HIV/AIDS, malaria, and tuberculosis in the developing world and is funded by taxes on airline tickets (Medicines Patent Pool, n.d.). The agency acts as an independent pool administrator, encouraging pharmaceutical companies to place their patents in the UNITAID pool. The aim of the pool is threefold:

- To bring the prices of medicines down.
- To facilitate the manufacture of single-dose medicines. HIV/AIDS is a condition that requires the use of a combination of different active ingredients and so requires the rights of different patent holders for the different active ingredients to be combined in order to produce single-tablet formulations (particularly important for the treatment of children but also in making the treatment as easy as possible).

- To help stimulate research into new paediatric formulations of the medicine. Although many children are infected with HIV/AIDS in the developing world, research in this area is limited because the bulk of HIV/AIDS research occurs in the developed world, where children generally do not get infected with HIV/AIDS as better treatment to prevent mother-to-child transmission is generally available.

HIV/AIDS is an example where pooling of pharmaceutical patents could actually be in the patent holders' commercial interests and in the interests of those with the disease as, without pools, a single-dose tablet cannot be produced (Bristows, 2009; Medicines Patent Pool, n.d.).

Another example of patent pooling with a social purpose is the Eco-Patent Commons initiative, which works on the understanding that businesses and academic institutions may hold some patents that provide environmental benefit but don't represent an essential core part of their business, and so may be willing to 'forego royalties when ecological use of the technology can improve the physical, economic, and business environment in which the company operates' (Eco-Patent Commons, 2015). The objectives of this patent pooling are:

- to provide an avenue by which innovations and solutions may be easily shared to accelerate and facilitate implementation to protect the environment and perhaps lead to further innovation;
- to promote and encourage cooperation and collaboration between businesses that pledge patents and potential users to foster further joint innovations and the advancement and development of solutions that benefit the environment (Eco-Patent Commons, 2015).

Since the launch of the Eco-Patent Commons, 105 eco-friendly patents have been contributed by 11 companies worldwide representing a variety of industries, namely Bosch, Dow, Fuji Xerox, Hewlett-Packard, IBM, Nokia, Pitney Bowes, Ricoh, Sony, Taisei, and Xerox, in addition to the hosting organization Environmental Law Institute (Awad, 2015).

Finally, under the Sustainable Development Goals process, the UN launched a new Technology Bank and Technology Facilitation Mechanism in October 2015. The Technology Facility Mechanism expects to start by supporting technology needs assessments in developing countries and to strengthen national institutional capacities – welcome but hardly radical moves (Casey, 2015). The Technology Bank, headquartered in Turkey, will have three functions:

- a patents bank to help developing countries secure relevant patents at negotiated or concessionary rates and to help protect IPR derived by least developed countries (LDC) inventors;
- a science and technology research depository to help countries access scientific literature, broker research collaboration through partnerships, and build capacity to expand the publication of scientific work from developing countries;

- a science, technology, and innovation supporting mechanism to build human and institutional capacity in the area of science and technology, establish technology incubators and ICT connectivity, market research results, and lever diaspora knowledge networks (Rahman, 2014).

The Technology Bank is not yet fully functional and so it is too early to judge its impact; however, some of the more radical proposals made at earlier stages seem not to have made it through to the current plan. It sets out some very modest ambitions with respect to IPR. Although the original proposal suggested the Bank should: 'Pragmatically address IPR with a balanced approach towards safeguarding the interests of LDCs and the technology holders including by exploring innovative approaches' (UN-OHRLLS, 2013), by the time the detailed feasibility study report was published two years later, this had morphed into a pretty much business-as-normal commitment that:

> The IP Bank should help build domestic capacities to absorb transferred patented-IP to LDCs. It should act as a conduit between IP rights holders in developed economies and relevant actors in the LDCs. The IP Bank should support negotiated agreements by providing expertise … to LDC participants, while ensuring that the respective interests of all parties are reconciled. It would be entirely voluntary and use conventional licensing of existing or expired patents and know-how and other knowledge (such as access to training, manuals, supply chain for purchase or donation of parts, etc.). (UN, 2015: 7)

Where next?

There are widespread market failures across multiple sectors with respect to technological innovation delivering either a universal social foundation or preventing our breaching planetary ecological boundaries. The patent system is part and parcel of that failure and, as has been noted, is not fit for purpose in terms of driving sustainable and equitable development. Stiglitz may be right not to advocate the abandonment of patents entirely, but the parasitic aspects of trolls and patent thickets need to be purged from the system and the TRIPS agreements rebalanced to reflect an IPR system that is in the interests of developing economies and sustainable development. The new UN Technology Bank and Technology Facilitation Mechanism could have been designed to help deliver some of these radical changes, but initial indications are that it will not challenge the status quo on intellectual property rights.

The alternatives to patents examined here all have their limitations. Patent pools may release some additional innovative momentum but there is a very limited number with a developmental purpose. Although it could be argued that the UNITAID pool for HIV/AIDS drugs provides a commercial driver for pharmaceutical companies to innovate, others, such as the Eco-Patent Commons, remain closer to corporate social responsibility exercises than truly transformative vehicles. While there is clearly a role for prizes and grants in shaping the direction of research and development, they have limitations and, for fields such as medicine (where the costs of bringing a novel and useful drug to commercial availability are enormous), would require a massive scale-up to be effective.

So what is the IPR system that will deliver the right drivers to push technology innovation in the direction we need it to go – towards solutions to the issues of poverty, social cohesion, justice, and environmental sustainability? The obvious answer, essentially the one Stiglitz provides, is a rebalanced 'portfolio' of all the above, combined with a clean-up of patents regulation. But it is difficult to see that as being enough, given the magnitude of the challenge we face and the relatively short amount of time we have to find solutions to the problem of avoiding irreversible climate change, as but one example. We are not looking for incremental change but normative change and this will require a more radical response. As far as drivers for technology innovation are concerned, this may mean drawing inspiration from the open-source movement rather than IPR legislation to forge a new path, something that will be returned to in Chapter 11.

References

Awad, B. (2015) 'Patent pledges in green technology (draft paper)', Program on Information Justice and Intellectual Property, Washington College of Law, Washington, DC: American University.

Boldrin, M. and Levine, D. (2013) 'The case against patents', *The Journal of Economic Perspectives* 27: 3–22 <http://dx.doi.org/10.1257/jep.27.1.3>.

Bristows (2009) 'The pros and cons of patent pooling' [online], <http://www.bristows.com/articles/the-pros-and-cons-of-patent-pooling> [accessed 14 March 2016].

Carter, Z. (2012) 'U.S. trade position protecting high drug prices blasted by U.N. agencies' [online], 6 January, <http://www.huffingtonpost.com/2012/06/01/us-trade-drug-prices-un_n_1560481.html> [accessed 10 November 2015].

Casey, J. (2015) 'UN places technology at centre of development' [online], 26 September, *Practical Action,* <http://practicalaction.org/blog/news/un-places-technology-at-the-centre-of-development/> [accessed 11 November 2015].

Eco- Patent Commons (2015) 'About Eco-Patent Commons' [online], *Eco-Patent Commons,* <https://ecopatentcommons.org/about-eco-patent-commons> [accessed 11 November 2015].

Fagerberg, J., Srholec, M., and Verspagen, B. (2009) 'Innovation and economic development', UNU-Merit Working Paper Series, Maastricht Economic and Social Research and Training Centre on Innovation and Technology, Maastricht: United Nations University.

Henry, C. and Stiglitz, J. (2010) 'Intellectual property, dissemination of innovation and sustainable development', *Global Policy,* 1(3): 237–51 <http://dx.doi.org/10.1111/j.1758-5899.2010.00048.x>.

Kaggle (2015) 'Kaggle competitions', <https://www.kaggle.com/solutions/competitions> [accessed 26 November 2015].

Kenny, C. and Barder, O. (2015) 'Technology, development, and the post-2015 settlement', CGD Policy Paper 63, Washington, DC: Centre for Global Development.

Khor, M. (2002) *Intellectual Property, Biodiversity and Sustainable Development – Resolving the Difficult Issues,* London: Zed Books.

Medicines Patent Pool (no date) <http://www.medicinespatentpool.org/about/> [accessed 14 March 2016].

Mazzucato, M. (2013) *The Entrepreneurial State: Debunking Public vs. Private Sector Myths,* London: Anthem Press.

Musungu, S.F. and Oh, C. (2005) *The Use of Flexibilities in TRIPS by Developing Countries: Can they Promote Access to Medicines?* Geneva: Commission on Intellectual Property rights, Innovation and Public Health, WHO.

Rahman, K. (2014) 'Technology Bank for LDCs', presented at UN-OHRLLS, 10 October, <http://www.un.org/esa/ffd/wp-content/uploads/2014/12/10Dec14-Rahman-Presentation.pdf> [accessed 11 November 2015].

Smith, J. (2009) *Science and Technology for Development,* London: Zed Books.

Stiglitz, J. (2008) 'Economic foundations of intellectual property rights', *Duke Law Journal,* 57: 1693–1724 <http://www.jstor.org/stable/40040630>.

UN (2015) *Feasibility Study for a United Nations Technology Bank for the Least Developed Countries,* New York: United Nations.

UN-OHRLLS (2013) 'A technology bank and science, technology and innovation supporting mechanism for the least developed countries', informal background note, New York: UN.

WIPO (2015) 'Frequently asked questions: patents', *World Intellectual Property Organization,* <http://www.wipo.int/patents/en/faq_patents.html> [accessed 10 November 2015].

CHAPTER 10

Recognizing the role of the state in effective innovation systems

Neoliberal orthodoxy maintains the role of government is to provide the 'light touch' regulatory environment that supposedly avoids market failures and allows markets to operate at their optimal efficiency in terms of allocating resources. Championed in the 1980s by the likes of Ronald Reagan ('Government is not the solution to our problem; government is the problem' (1981)) and Margaret Thatcher ('There can be no liberty unless there is economic liberty' (1979)), this has become an accepted view of the distinction between private and public-sector roles. How that conventional view places the government's role in supporting technology innovation was well summarized by *The Economist* in an article on the digitalization of manufacturing:

> Consumers will have little difficulty adapting to the new age of better products, swiftly delivered. Governments, however, may find it harder. Their instinct is to protect industries and companies that already exist, not the upstarts that would destroy them. They shower old factories with subsidies and bully bosses who want to move production abroad. They spend billions backing the new technologies which they, in their wisdom, think will prevail ... None of this makes sense ... Governments have always been lousy at picking winners, and they are likely to become more so, as legions of entrepreneurs and tinkerers swap designs online, turn them into products at home and market them globally from a garage. As the revolution rages, governments should stick to the basics: better schools for a skilled workforce, clear rules and a level playing field for enterprises of all kinds. Leave the rest to the revolutionaries. (*The Economist,* 2012)

As this book has shown, however, there is a wide range of vitally important issues the 'revolutionaries' have yet to tackle. Market forces are not driving technological innovation at anywhere near the speed or volume required to address the problems of access to basic services for the poor in the developing world or of creating an environmentally sustainable global economy. Moreover, these market failures are of a scale and nature that will take more than the removal of red tape and a few regulatory barriers to fix. Indeed, in many of these cases, delivering the technologies at scale will involve levels of risk that the private sector cannot take on alone and activity that the private sector is not equipped to drive – in particular, the nurturing of the breadth of institutions and relationships necessary to create successful national innovation systems.

In reality, governments (even neoliberal ones) recognize this to an extent. As discussed in Chapter 8, views on the role that governments can play in

http://dx.doi.org/10.3362/9781780449043.011

technology development have been evolving in the innovation systems literature. A recent publication by the UK government, for example, notes that the globalization of markets does not imply that national governments are powerless to intervene: 'Decisions on science facilities and performance, education, the regulatory framework, and above all knowledge and information infrastructures remain open to discretionary commitments by national governments' (Department for Business Innovation and Skills, 2011: 22). Some analyses go further to argue that government should, and often in reality does, play an active and entrepreneurial role itself, a role that goes beyond correcting market imperfections and making sure the education system works to being directly engaged in the shaping of economies. This chapter draws heavily on the work of Mariana Mazzucato, RM Philips Professor in the Economics of Innovation at the Science Policy Research Unit, University of Sussex, to look at that idea in more detail.

Venture capital and the valley of death in the energy sector

The 'valley of death' is a term typically used to describe a particular part of the journey from a prototype technology developed through research and development to full commercialization and scale-up. It is a point at which capital requirements multiply rapidly and the initial grant finance, seed money, or small business loans run out, but when risks are perceived to be too high for banks – the common source of project funding for scale-up and commercialization – to be interested in stepping in. Many technology businesses fail to cross the valley of death and so, as one writer puts it, 'the business runs out of cash and out of steam and dies a painful death in a landscape littered with the carcasses of companies that came before' (Clements, 2011).

Clean power-generation technology in the US

In considering the clean-energy technology sector, views vary as to whether the process of laboratory to commercial scale-up has four stages with one valley of death (Gosh and Nanda, 2010) or five stages with two valleys of death (Jenkins and Mansur, 2011), but essentially the issue remains the same – the existence of one or two finance gaps that act as a brake on the rapid commercialization of new clean-energy technology. Figure 10.1 illustrates the cycle. In a developed-country economy, such as the US, traditional venture capital is the source of funding sought to bridge 'valley of death' gaps. Venture capital funds specialize in high-risk situations, but to do this they are generally structured to make a relatively large number of comparatively small investments, knowing that only a small proportion will generate good returns, thus spreading risk as a mitigation strategy. Gosh and Nanda provide a breakdown of a typical venture capital fund's portfolio, based on a series of interviews with funds, as shown in Table 10.1. Typical investments from alternatives such as angel investors, who take a share of equity in return for an investment, are often even lower, at $1–2 mn per investment.

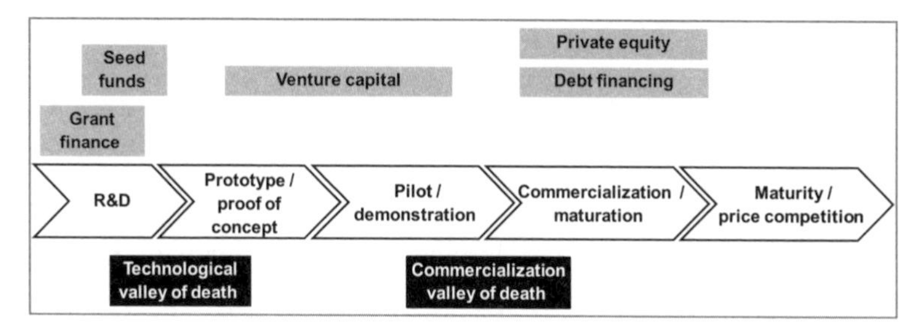

Figure 10.1 The clean-energy innovation cycle and valleys of death

Source: based on Jenkins and Mansur, 2011: 5

Table 10.1 Investment and targeted returns for a typical US venture capital fund portfolio

Outcome	*Investment per company ($ m)*	*Expected no. of investments*	*Total invested ($ m)*	*Total return ($ m)*
Early failure	$5	5	$25	0
Complete write-off	$8–15	5	$55	0
Money back	$8–15	5	$55	$50
Successful exit (low)	$8–15	5	$55	$200
Successful exit (medium)	$8–15	5	$55	$350
Successful exit (high)	$8–15	5	$55	$500
Total			*$300*	*$1,100*

Source: based on Ghosh and Nanda, 2010: Figure 2

The limited scale of typical venture capital or angel investments poses a big problem for the development and commercialization of new clean-energy technology. As a Breakthrough Institute report on valleys of death in the sector in the US notes:

> The early stage expenses necessary for nascent advanced energy technologies to demonstrate market validity, including prototyping and laboratory costs, are significantly higher than many other sectors. In the 'garage culture' of internet start-ups, for example, it takes comparatively little capital or time to advance an innovative research idea or product concept into a provable business plan. In contrast, bringing innovative energy research to its pre-deployment phase requires significant capital and as much as 10–15 years' time'. (Jenkins and Mansur, 2011: 7)

The time factor is as much an issue for venture capital funds as the investment requirement. Although venture capital funds are generally built to have a life of 10 years, the way the incentives for fund managers are structured (typically an annual management fee of 2 per cent of the capital committed to the fund plus a bonus for exceeding the targeted return on the fund that could be as much as 20 per cent of the excess return) encourages early exit, both to harvest any excess return as soon as commercial viability is proven and to establish a track record with their own investors in order to raise funds for the next investment. This acts to focus venture capital funds on opportunities that have the best chance of achieving commercial viability within three to five years, too short a time scale for many new power-generation technologies to become established (Gosh and Nanda, 2010).

Not all new energy technology development faces this financing barrier, as Nanda and Gosh make clear when they map technologies onto potential funding sources (see Figure 10.2). But for the development and deployment of new clean power-generation technology, where major investments of the order of hundreds of millions of pounds, dollars, or euros may be required over time scales of 10–15 years, a 'valley of death' clearly exists, which needs to be addressed urgently.

There are several interventions governments can make beyond the traditionally ascribed role of grant financing basic research. These typically fall into two categories. Firstly, governments can try to provide financing themselves to bridge the valley of death directly. In the US, the Advanced Research Projects Agency – Energy programme (ARPA-E) is one such example, investing between $250 mn and $300 mn a year in energy-related technology for transportation and stationary power systems since 2013 (ARPA-E, 2015).

Secondly, they can try to create the conditions that would encourage private-sector finance to flow into these riskier investments. A range of

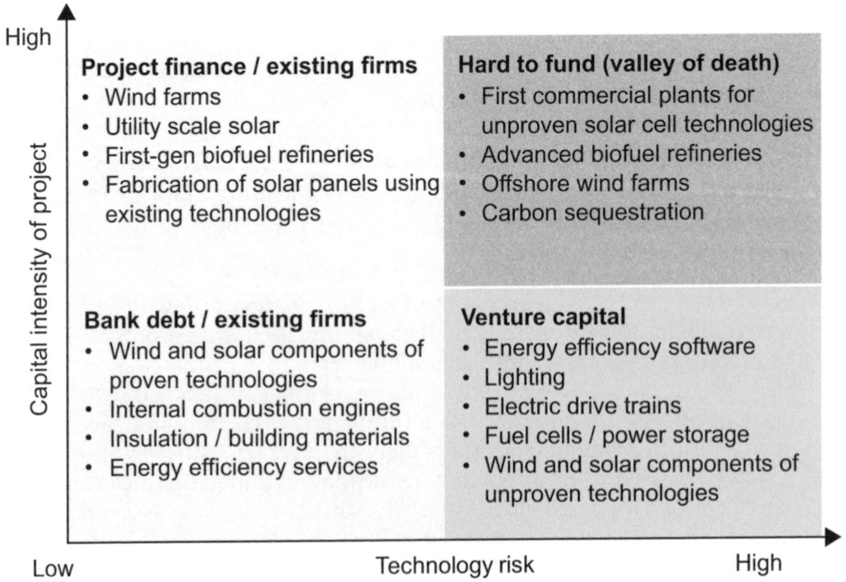

High
Capital intensity of project

Project finance / existing firms
- Wind farms
- Utility scale solar
- First-gen biofuel refineries
- Fabrication of solar panels using existing technologies

Hard to fund (valley of death)
- First commercial plants for unproven solar cell technologies
- Advanced biofuel refineries
- Offshore wind farms
- Carbon sequestration

Bank debt / existing firms
- Wind and solar components of proven technologies
- Internal combustion engines
- Insulation / building materials
- Energy efficiency services

Venture capital
- Energy efficiency software
- Lighting
- Electric drive trains
- Fuel cells / power storage
- Wind and solar components of unproven technologies

Low Technology risk High

Figure 10.2 Energy-sector technology innovation projects and types of finance available

Source: based on Gosh and Nanda, 2010: Figures 4 and 5 (reproduced with permission of Harvard Business School. Copyright 2014 by the President and Fellows of Harvard College; all rights reserved.)

instruments has been used to do this in the past. Some of these have focused on stimulating demand for renewable energy. Here there are problems that are peculiar to the energy sector. A new drug or a new piece of software will compete in a marketplace based not just on price but on the new features they offer, conferring intrinsic value themselves. Energy, on the other hand, is a commodity that is used to deliver other services. From the energy consumers' perspective, energy from a renewable source offers no additional feature or value over and above that generated from fossil fuels and so competes with incumbent fossil fuel-based generation on price alone. Although things are slowly changing, fossil fuel power generation has traditionally been cheaper than renewables in many instances, hence the lack of a price signal to attract new investors into the market at the scale needed.

Feed-in tariffs (FITs) have been one way in which governments have tried to counter this problem. By offering a guaranteed and often higher than market rate to purchase renewable power for a fixed and substantial period of time (typically up to 25 years), governments can provide both the price signal and a signal around the long-term stability of returns that can take risk out of the environment for potential private investors. A handful of countries have tried this, of which Germany probably exhibits the greatest success in both stimulating technology innovation and delivering a greater share of generating capacity from renewable sources. Not all attempts have been a success, though: the Spanish government overstretched itself and had to back out of its offer, while the UK government's inconsistent use of FITs has caused great uncertainty in the national market.

Another approach to creating demand uses the purchasing power of government, often the largest single buyer of power, to set trends:

> Examples abound of governments around the globe taking a direct role in fostering clean energy technologies. In the UK, the quasi-governmental Crown Estate has agreed to purchase the first 7.5 MW Clipper wind turbine when it is complete in two to three years. In Brazil, state-owned utility Electrobras has guaranteed 20-year clean energy power purchase agreements totalling 3,300 MW and resells the power to distributors. In China, the central government has issued a plan to add more than 60,000 energy-saving or new energy government-owned vehicles by 2012. (Bloomberg New Energy Finance, 2010)

Other proposals include using regulation or corporate incentives to encourage utilities to become first adopters of new technologies (Gosh and Nanda, 2010) and new public–private co-investing models which allow the public sector to choose the technologies to support and the private sector to buy out the public share of the investment if the technology turns out to be commercially viable (Bloomberg New Energy Finance, 2010). The setting of a suitably high enough carbon price would obviously be another positive price signal.

Off-grid lighting in the developing world

Death valleys exist in other sectors, too, where weak markets fail to deliver the signals needed to drive investment into socially or environmentally useful technology development. The failure of markets to deliver new drugs to treat diseases that predominantly affect populations of developing countries is another example. This is covered in Chapter 8, which describes advanced market commitments (AMCs) among other efforts to correct those market failures. AMCs are, in essence, simply another example of a necessary state-driven response to bridging a 'valley of death'.

Some valleys are not quite as wide or deep as others. The investments required to deploy new off-grid technologies to meet energy access demand for rural populations in the developing world are orders of magnitude smaller, and with far shorter potential time scales from R&D to commercialization, than the development of new drugs or utility-scale power-generation technology. But though the valleys may be smaller, they are just as effective at preventing large-scale deployment of socially and environmentally useful technologies.

Recent developments in the solar off-grid lighting product sector have focused on making products affordable to 'bottom of the pyramid' consumers in developing countries by combining the latest technology with new financing models. Low-cost solar photovoltaic panels, next-generation lithium batteries, low-power light-emitting diode (LED) lights, and charging facilities for mobile phones are being combined to provide small-scale solar home systems for prices typically around the $200 mark. Pay-as-you-go solutions, such as M-KOPA's in Uganda and Kenya, allow the cost of owning this technology to be spread through daily or weekly payments over a year to 18 months (MKOPA, 2014) or, as is the case with Off Grid Electric in Tanzania, accessed via a leasing system (Off Grid Electric, 2012). In either case, the approach brings the cost of access to electric lighting down to roughly 45 cents a day, which compares favourably with typical household costs of 50–60c per day for kerosene for lighting plus typical mobile charging costs that families would be paying for anyway.

M-KOPA has partnered with the mobile network Safaricom in Uganda and Kenya to enable its customers to use the M-Pesa mobile money system and to make its products widely available through the Safaricom network of shops. Customers make a deposit payment to take home a solar system and then activate it using a code accessed by purchasing a widely available scratch card. Regular payments are made using a mobile money platform via the embedded SIM card. Off Grid Electric's leasing model doesn't embed a SIM in the solar home system but allows customers to purchase activation codes from their own phones or by purchasing further scratch cards. In the case of non-payment, systems can be turned off remotely by the distributor where there is an embedded SIM card in the solar home system. Alternatively, in the scratch card approach, the codes from each card purchased are time limited, after which the system switches off and requires a new scratch card to be purchased for a new code to reactivate.

These models of providing distributed off-grid access to electricity are showing real promise, buoyed by the fact that the viability of solar systems to provide domestic power has never been greater. The cost of solar panels has plummeted in recent years, falling 75 per cent since 2009 (Jenkins, 2015) while, at the same time, lithium batteries have become far more efficient vehicles for storing power. Perhaps even more importantly, the power requirements of appliances have also dropped with the advent of not just LED lighting but also super-efficient fridges, televisions, and fans, meaning you can get much more out of a solar panel than was the case even two years ago. The same 40-watt solar panel that 10 years ago could power one 25-watt light bulb can today power four LED lights, a phone charger, a radio, and maybe even a small colour TV. The size of the demand in sub-Saharan Africa (600 million people without electricity) and the falling price of the technology, combined with pay-as-you-go approaches to consumer financing, have created market-based opportunities to tackle the energy access challenge using these solar home systems and lights.

There are signs of the potential of this model to go to scale. For example, the social enterprise Sunny Money has sold over 1.7 million solar lamps in Tanzania, Kenya, Uganda, Malawi, and Zambia to date, while M-KOPA's customer base grew rapidly from 60,000 customers in April 2014 to 250,000 by the end of 2015 and Off Grid Electric added 10,000 customers a month in 2015. Indeed, these initial signs have led organizations such as Power for All to claim that universal access to energy could be achieved as early as 2025 and at a cost of just $70 bn, which makes for a far brighter outlook than the generally accepted prediction by the International Energy Agency in 2011 of a completion date of 2035 contingent on an expenditure of $700 bn (Tice and Skierka, 2014).

The nature and capital intensity of the technology involved should situate the off-grid distributed energy sector in developing countries well inside the bottom right quadrant of Figure 10.2 and so in the right territory for venture capital funding. But the sector has its own valley of death. A study of investment needs in the off-grid lighting sector confirmed that access to working capital or long-term growth financing remains a major barrier to the development of the industry and its potential to scale up (AT Kearney, 2014). Among the reasons why investors are put off investing, the report cites:

- currency risks for external financing;
- uncertain legal and policy frameworks, for example, around whether import duties and value-added tax are levied on renewable energy equipment (which can lead to uncompetitive pricing compared to kerosene lighting, especially where kerosene is subsidized) or the existence and enforcement of quality standards (there is some evidence of market spoilage where poor-quality solar products have led to a loss of consumer confidence);
- limited track record of existing distributors;
- lack of awareness of solar solutions by consumers;

- cost of last mile distribution systems;
- limited access to consumer credit to make systems affordable.

There have already been state and multilateral agency-funded initiatives to address some of these problems. The Lighting Global Programme of the International Finance Corporation (IFC) and the World Bank, together with the United Nations Environment Programme and the German government development agency GiZ, have worked together to encourage regulators in developing countries to waive duties and taxes for renewable energy equipment. They have also helped to develop technical specifications and standards for future off-grid lighting systems and mounted awareness campaigns among some off-grid populations. But much more support is required and the report lists a number of other areas where state intervention and financing will be vital to move things to scale. These include:

- working capital funds specifically aimed at off-grid projects, along the lines of the $10 mn fund already set up by the IFC;
- loan and export guarantees to reduce investment risk;
- assistance facilities to provide technical, commercial, and strategic advice to new players, to give investors assurance that the firms they may fund have the necessary capabilities to manage their finances and to avoid default;
- industry-specific, public–private, small and medium-sized, enterprise growth capital funds – an approach that has been shown to work in the biotechnology industry, where public funding is used to lever private investment and risks are shared;
- the creation of sharing and learning networks for investors to provide market information and access to peer investors to reduce due diligence and transaction costs.

As is the case for the introduction of new large-scale power-generation technology in the US or the development and deployment of new drugs for neglected diseases, the further development and widespread deployment of off-grid distributed solutions to energy poverty will not progress without concerted action from state and state-funded institutions that goes well beyond light-touch regulation and involves an active engagement with, and shaping and supporting of, a new market.

Recognizing reality: governments engage in entrepreneurial activity

State involvement in the creation of general-purpose technologies

The previous section highlighted the need for state finance and for deliberate and extensive state action to de-risk the environment and to create the conditions under which private capital might, often alongside public finance, flow into the sector to bridge the 'valley'. In reality, state engagement in entrepreneurial development of new technologies in areas of high risk or high complexity is far more common than often imagined and goes far beyond simply bridging the financial valley of death.

In Chapter 8, it was noted that big pharmaceutical companies in the US had a poor track record of investing in new drugs as opposed to 'me too' variants of existing ones. Of the 146 out of 1,072 approved drugs classified as new molecular entities (NMEs) or completely new medicines from 1993 to 2004, 75 per cent could be traced back not to private corporations, but to the publicly funded National Institute of Health's laboratories (Mazzucato, 2013: 66). Mazzucato compares this reality to a 2010 statement from Andrew Witty, then CEO of GlaxoSmithKline, to the effect that the pharmaceutical industry was hugely innovative and that if governments were only to work together to encourage rather than stifle innovation, the industry could deliver the next era of 'revolutionary medicine'. She notes wryly that 'it is the revolutionary spirit of the State labs, producing 75 per cent of the radical new drugs, that is allowing Witty and his fellow CEOs to spend most of their time focusing on how to boost their stock prices (for example, through stock repurchase programmes)' (Mazzucato, 2013: p 67).

Mazzucato goes on to show how a whole range of general-purpose technologies (GPTs), which have themselves formed platforms for further technological innovation, came into being primarily through government-funded and government-led research and development. GPTs are defined as having three important characteristics:

- They are pervasive in that they are adopted and adapted across many sectors.
- They continue to improve over time, reducing in cost as a consequence.
- They provide a platform that can make it easier for further innovation and the invention of other new products (Mazzucato, 2013).

Mazzucato uses work by Vernon Ruttan (2006) to argue that large-scale and long-term government investments have been behind most of the GPTs developed over the last 100 years, including aviation technologies, space technologies, computers and semi-conductor technologies, the internet, and nuclear power. She adds nanotechnology to this list. The example she gives of Apple's iPod is particularly illustrative of how a company that is known as a technology innovator is, in fact, commercially rather than technically innovative, establishing its niche by the inventive recombination of existing and largely publicly funded technology R&D products. Figure 10.3 illustrates the dependency of the iPhone family not on Apple's own technology research, but on research and product development from a range of US and other government institutions.

All the technologies that make the iPhone and its siblings 'smart' have their origins in government-funded programmes, whether that is the internet, global positioning system (GPS) technology, the touchscreen display, or even

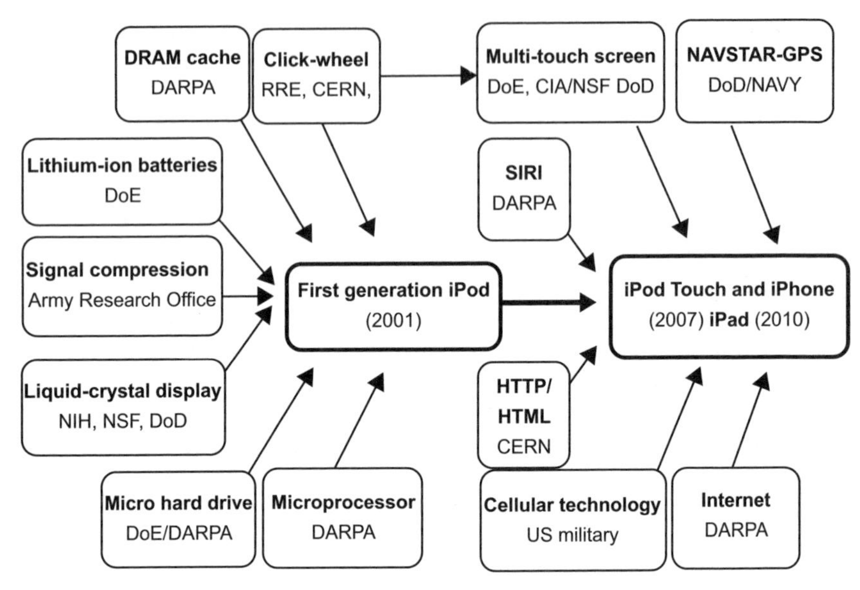

Figure 10.3 The origins of technology used in Apple's iPod, iPhone, and iPad

Note: CERN = European organization for nuclear research, CIA = Central Intelligence Agency (US), DARPA = US Defence Advanced Research Projects Agency, DoD = Department of Defence (US), DoE = Department of Energy (US), NIH = National Institutes of Health (US), NSF = National Science Foundation (US), RRE = Royal Radar Establishment (UK)

Source: Mazzucato, 2013: 109

the artificial intelligence behind the latest voice-activated SIRI personal assistant.

State involvement in the development of GPTs suggests a very different role from the one prescribed in the quote from *The Economist* at the beginning of this chapter. Rather than 'sticking to the basics' of better schools for a skilled workforce and the development of clear rules and a level playing field for enterprises of all kinds, this represents the state as an entrepreneurial actor, a driver of the technological revolution itself. Indeed, in highly uncertain or complex situations, where clear market signals are absent but progress is essential, the state may be the only institution able to bear the risks involved in bringing a new technology to maturity, creating the new GPTs which the private sector can then use to combine and innovate further, technically and commercially. Using clean energy as an example, Mazzucato argues that 'it is not just about the willingness of the state to lead, but the willingness to *sustain* support for new and transitional technologies until industry can "mature" – until the cost and performance meet or exceed those of incumbent technologies (for example, fossil power)' (Mazzucato, 2013: 196). This goes beyond using a few incentives or minor regulatory fixes to nudge markets in the right direction. It is about choosing a direction and providing a prolonged push to actively get things started.

Changing the narrative: rebalancing expectations of public and private-sector roles

So what does all this mean for the governance and oversight of technology innovation? A Technology Justice view would be that the roles of the public and private sectors need to be recalibrated to ensure that, combined, they create an ecology of actors, policy, finance, and capacity that has, as its natural outcome, technology innovation which is focused on the problems of poverty and environmental sustainability. Some form of transformation is required to achieve this balance. Drawing on the work of Mazzucato (2014), there are three important strands of that transformation.

From lousy pickers of winners to active entrepreneurial risk-takers

The public narrative needs to change. The myth referred to in *The Economist* quote, that 'Governments have always been lousy at picking winners', has to be debunked on two counts. Firstly, the state does have, and frequently exercises, an entrepreneurial role, usually in areas where risks or complexities are too high for the private sector to act alone. Public policy needs to recognize and welcome this and actively build state capabilities to fulfil the entrepreneurial role effectively – a real challenge in an era where, in Europe and elsewhere, government expenditure is being slashed under austerity policies in response to the post-2008 economic crisis. Secondly, the entrepreneurial role of the state is to set direction where market signals are not strong enough (for example, for clean energy, medical research for orphan diseases, or agroecological approaches to food production), then not to pick one winner but to invest in *many* potential winners, with the same logic that venture capital funds apply to expectations of success and failure rates. If the state must be the prime investor in areas of the highest risk, it must have permission to make investments that fail in order to find ones that will succeed.

From parasitic to symbiotic public–private partnerships

In addition to changing expectations on the role the state should play in the innovation process, we need to raise expectations concerning the contribution the private sector will make. Some of the examples in previous chapters demonstrate there is now a somewhat parasitic relationship between the private sector and the state in some fields of technology innovation. That Apple relies on the products of state-funded technology innovation to develop new products or that drug companies rely on state-funded laboratories to develop the NMEs they subsequently exploit is not in itself parasitic. But when transnational companies privatize the benefits from public investment in technology R&D while, at the same time, doing everything possible to minimize their liability for tax in the domicile where that original value was created, then the relationship between private and public sectors becomes parasitic.

Added to this is the problem Mazzucato refers to as the 'financialization' of the private sector, which diverts attention away from genuine innovation to create value through new products and towards activities that involve productless value extraction instead. As mentioned, one example is that of venture capital funds, which are generally structured in a way that incentivizes fund managers to extract value, sell up, and move on as quickly as possible, making it harder for those funds to be invested in technology innovation for difficult markets that will produce social or environmental returns. Another example Mazzucato uses is that of big companies reducing spending on R&D while simultaneously increasing the amount of profit they reinvest in buying back their own shares, a strategy that is used to boost share price, which in turn impacts on the value of the stock options that form part of senior executive remuneration packages. In 2011, Pfizer, for example, used 90 per cent of its profit to pay out $6.3 bn in dividends and repurchase $9 bn of its own stock, an expenditure roughly equal to its entire R&D spend for the year. Likewise, the biotechnology company Amgen engaged in stock repurchases equivalent to 115 per cent of R&D expenditure over the period 1992–2011 (Mazzucato, 2013). Meanwhile, in 2010, the US American Energy Innovation Council asked the US government to treble its spend on clean technology and provide an additional $1 bn to ARPA-E, despite the fact that seven of the companies that formed the council together had sufficient resources to spend $237 bn on stock repurchases between 2001 and 2010 (Mazzucato, 2014).

The private sector and private capital unquestionably have an important role to play in national innovation systems and in driving technology innovation towards social and environmental goals. But we have to find a new set of drivers, most likely by reforms to tax systems and putting limits on activities such as share buy-backs, that will limit parasitic, risk-avoiding, and value-extractive behaviour by corporations and instead encourage collaborative and risk-sharing joint ventures between private companies and the state.

From de-risking for private investors to the sharing of risk and reward

A third part of the great transformation needed relates to Zeigler's idea of a fair space for technology innovation, discussed in Chapter 7: the sense that in order for technology innovation to be fair, we not only have to consider whether the distribution of its resulting impact is just, but also whether the gains arising from innovation are equitably distributed according to the public and private investments that were necessary to allow it to happen. In short, Zeigler uses Mazzucato's work to argue that:

> if the state contributes via its interventions to the process of creating major science-based innovations (such as the internet or renewable energies), especially if the state thereby takes long-term risks that private actors are unlikely to take – both in terms of willingness and also in terms of capacity – then the state should also 'reap some of the financial rewards' from these investments. (Ziegler, 2015)

This reward sharing could take different forms. For example, the state could take a 'golden share' in patents, which would allow it not only to benefit from the proceeds of licensing but also give it a say in what further use the patent is put to. Or it could retain some equity in companies and either accept dividends or seek a percentage of the returns made from the utilization of the technology innovation it invested in. Financial returns to the state from shares in patents or equity stakes in companies could flow into some sort of national investment fund that, in turn, could be used for further investments in other technology innovation efforts. Allowing the state to reap a return on high-risk investments in technology innovation would have the double benefit of creating a new source of national investment funds at a time when government budgets are under strain, while allowing the tax-paying public to see a clearer link between public expenditure and return, building political support for much-needed investment (Mazzucato, 2013: 197).

In conclusion, and to be absolutely clear, there are important roles for both the private and public sectors to play in ensuring technological innovation addresses crucial environmental and poverty goals. But time is limited and the pervasive narrative that the role of government is primarily to regulate markets has to be challenged in order to make faster progress. As the analysis in this chapter has shown, we do not have the optimum balance necessary to deliver desperately needed technological innovation. Mazzucato's three action points – recognizing the role of governments as entrepreneurial risk-takers, tackling parasitic and risk-avoiding behaviour arising from the financialization of the private sector, and ensuring a fair return on public investment – would go a long way towards redressing the current imbalance.

References

ARPA-E (2015) *2015 Congressional budget – Advanced Research Projects Agency – Energy (ARPA-E),* Washington, DC: US Government Advanced Research Projects Agency – Energy.

AT Kearney (2014) 'Investment and finance study for off-grid lighting', Utrecht: Global Off-Grid Lighting Association.

Bloomberg New Energy Finance (2010) *Crossing the Valley of Death: Solutions to the Next Generation Clean Energy Finance Gap,* London: Bloomberg New Energy Finance.

Clements, E. (2011) 'Crossing the valley of death' [online], 1 February, *Symmetry Magazine* <http://www.symmetrymagazine.org/article/february-2011/crossing-the-valley-of-death> [accessed 14 November 2015].

Department for Business, Innovation and Skills (2011) *Innovation and Research Strategy for Growth,* London: UK Government.

The Economist (2012) 'The third industrial revolution', *The Economist,* 21 April <http://www.economist.com/node/21553017> [accessed 17 March 2016].

Nanda, R. and Ghosh, S. (2010) 'Venture capital investment in the clean energy sector', Harvard Business School Case 814052.

Jenkins, S. (2015) 'Falling cost of solar offers solace after halving of oil price', *Financial Times*, 31 March, <http://www.ft.com/cms/s/0/ee666260-d149-11e4-86c8-00144feab7de.html#axzz3reQjA2bB> [accessed 16 November 2015].

Jenkins, J. and Mansur, S. (2011) 'Bridging the clean energy valleys of death: helping American entrepreneurs meet the nation's energy innovation imperative', Oakland, CA: Breakthrough Institute.

Mazzucato, M. (2013) *The Entrepreneurial State: Debunking Public vs. Private Sector Myths,* London: Anthem Press.

Mazzucato, M. (2014) 'The (green) entrepreneurial state: risks, rewards, and directionality in innovation', 13 May, LinkedIn SlideShare, <http://www.slideshare.net/Stepscentre/mariana-mazzucato-the-green-entrepren> [accessed 16 November 2015].

M-KOPA (2014) 'Products', *M-KOPA Solar,* <http://www.m-kopa.com/products/> [accessed 16 November 2015].

Off Grid Electric (2012) 'Our story', *Off Grid Electric*, <http://offgrid-electric.com/#home> [accessed 16 November 2015].

Reagan, R. (1981) 'Inaugural address, January 20', *Ronald Regan Presidential Library & Museum*, <http://www.reagan.utexas.edu/archives/speeches/1981/12081a.htm> [accessed 16 November 2015].

Ruttan, V.W. (2006) *Is War Necessary for Economic Growth? Military Procurement and Technology Development*, Oxford: Oxford University Press.

Thatcher, M. (1979) 'World: An interview with Thatcher', 14 May, *Time Magazine,* <http://content.time.com/time/magazine/article/0,9171,916774,00.html> [accessed 13 November 2015].

Tice, D. and Skierka, K. (2014) 'The energy access imperative', Power for All <http://www.powerforall.org/resources/> [accessed 17 March 2016].

Ziegler, R. (2015) 'Justice and innovation – towards principles for creating a fair space for innovation', *Journal of Responsible Innovation*, 2(2): 184–200 <http://dx.doi.org/10.1080/23299460.2015.1057796>.

CHAPTER 11
Beyond market forces: other drivers for innovation

A recurring theme of the last four chapters of this book is the limitations of relying on market forces as the core driver of technological innovation to address environmental sustainability or poverty. The question therefore arises as to what alternative means exist that might help point technology innovation in the right direction? Some answers have been provided in the preceding chapters with regard to the role of government and the use of such mechanisms as prizes, grand challenges, and advanced market commitments in lieu of or alongside intellectual property rights. This chapter takes a different tack and considers three approaches drawn from answering three key questions: What is innovation for? Who is it for? And who does the innovating?

Innovation for what? Responsible research and innovation as a governance tool

A growing body of literature on the idea of responsible research and innovation (RRI)[1] may have relevance to Technology Justice and the challenge of governance of technology innovation. RRI shifts the emphasis from the management of risk (for example, via the precautionary principle) to a broader concern for the governance of the overall direction and purpose of innovation itself. It has arisen out of a growing recognition in the scientific community of the 'limitations of governance (of innovation) by market choice' and that, with innovation, the past and present don't always provide a good guide to the future, meaning we face a 'dilemma of control in that we lack the evidence on which to govern technologies before pathologies of path dependency and technological lock-in … set in' (Stilgoe et al., 2013: 1569). In short, RRI seeks to raise the question 'innovation for what?' into the process of technology innovation governance.

Stilgoe and colleagues (2013) offer a definition of RRI as the 'taking care of the future through collective stewardship of science and innovation of the present'. They provide a set of questions that have emerged as important during public debates on science and technology, questions that public groups want to see scientists and researchers asking themselves more often (see Table 11.1). The questions are grouped under three headings: product questions, process questions, and purpose questions. The authors note that conventional governance focuses mainly on the product questions and notions of technological risk, which can obscure areas of uncertainty about other, broader risks and about benefits. A focus on product questions alone,

http://dx.doi.org/10.3362/9781780449043.012

Table 11.1 Potential lines of questioning for responsible innovation governance approaches

Product questions	*Process questions*	*Purpose questions*
How will the risks and benefits be distributed? What other impacts can we anticipate? How might these change in the future? What don't we know about? What might we never know about?	How should standards be drawn up and applied? How should risks and benefits be defined and measured? Who is in control? Who is taking part? Who will take responsibility if things go wrong? How do we know we are right?	Why are researchers doing it? Are these motivations transparent and in the public interest? Who will benefit? What are they going to gain? What are the alternatives?

Note: based on Table 1 in Stilgoe et al., 2013: 1570

as has been argued elsewhere in this book, ignores important opportunities to ensure the overall purpose of research aligns with and advances knowledge around key societal challenges, hence the critical importance of including process and purpose questions in the proposed framework.

In response to this set of questions, the authors offer a possible responsible innovation governance framework based on four dimensions:

Anticipation. The prompting of 'what if?' questions to consider what is known, what is likely, what is plausible, and what is possible. Asking these questions also takes us beyond the potential risks and benefits associated with a technology to predicting and shaping desirable futures, while instilling some responsibility and ethics into what can sometimes lead to highly optimistic promises of major industrial and social transformation (see Grunwald (2004), for example, for further discussion of the use of visioning techniques to assess future technology scenarios).

Reflexivity. Institutional reflexivity involves holding up a mirror to a piece of R&D, its activities, commitments, and assumptions, and being aware of limitations of knowledge and that a particular framing of an issue may not be universally shared. 'Reflexivity directly challenges assumptions of scientific amorality and agnosticism [and] asks scientists, in public, to blur the boundary between their role responsibilities and wider, moral responsibilities' (Stilgoe et al., 2013: 1571).

Inclusion. As public concerns have grown around the potential environmental, health, and other impacts of rapidly evolving areas of science, such as nanotechnology and genomics, there has been a trend to explore how public engagement 'upstream' in the technology development process can bring societal influences to bear on the direction of innovation – before new technologies build momentum and choices become relatively locked in. The Danish Board of Technology's consensus conferences, which have been running since the 1980s, are one of the best known examples of such consultations (Fisher et al., 2006). Engagement with the public in conferences, citizens' juries, focus groups, and so on can also provide space and opportunity for reflexivity during R&D processes themselves.

Responsiveness. Anticipation, reflexivity, and inclusiveness are not useful as characteristics of responsible innovation unless there is also the capacity to change the shape or direction of research in response to changing circumstances or to the expression of societal values. Responsiveness thus requires researchers to respond to the political economy of not only the products but also the purposes of R&D and, for example, to ensure that research is informed by important societal challenges.

Applications of RRI

Responsible research and innovation has been attracting increasing attention in Europe and the UK since 2010 (see, for example, European Commission, 2012; Owen et al., 2012; Sutcliffe, 2015). The European Commission has published its own framework of six actions to deliver RRI which, to an extent, offers a practical realization of the four dimensions listed above:

1. *Engagement* to identify societal challenges via consultation with a wide range of societal actors, including industry, civil society, policymakers, and society;
2. *Gender equality,* both in terms of a balanced representation of men and women in the research workforce and in the focus and purpose of innovation;
3. *Science education* to enhance the current education process to better equip future researchers and other societal actors with the necessary knowledge and tools to fully participate and take responsibility in the research and innovation process;
4. *Open access* to make research and innovation both transparent and accessible, with free online access to the results of publicly funded research (publications and data);
5. *Ethical standards* to ensure that research and innovation respects fundamental rights and delivers increased societal relevance and acceptability of research and innovation outcomes;
6. *Governance* models for RRI that integrate the above five characteristics (European Commission, 2012).

RRI is embedded in a modest way in the EC's Horizon 2020 initiative – an €80 bn, seven-year (2014–20) funding programme to support 'smart, sustainable and inclusive growth and jobs'. Funding is available for research and innovation around seven 'grand societal challenges', including health, clean energy, climate, and sustainable agriculture, and for further work on RRI itself, including €462 mn to explore the six actions in the framework for implementing RRI (European Commission, 2013; 2015b).

Although interest in RRI has grown rapidly in Europe, and to an extent the US, little attention has been paid to its relevance in the context of developing countries, and its application in the developing world appears to be an under-researched area. The topic is also noticeably absent from documentation establishing the UN Technology Facilitation Mechanism for the Sustainable Development Goals or relevant material on science and technology innovation published under the Technology Mechanism of the United Nations Framework Convention on Climate Change.[2]

RRI as a tool for Technology Justice

Some iterations of RRI clearly have relevance to Technology Justice, notably the idea that we need a new system of governance for technology innovation that engages with wider society to look not just at risk but also the general purpose of science and technology development, debates around societal grand challenges, and how R&D agendas can be aligned with those challenges.

Applications of the concept to date are clearly limited, however. Although the EC's Horizon 2020 initiative – where the most practical work on RRI seems to have been done so far – claims that RRI is cross-cutting, it is difficult to find any reference to the application of the principles in, for example, the online manual for applying for research funds under the initiative (European Commission, 2015a). This, combined with the fact that no moves have yet been made to embed the principles in a set of binding EC standards means, at best, the principles are being considered only in the context of research funded by the EC itself and, even there, they are most likely operating as a 'nice to have' option as opposed to a fundamental requirement for the award of funding.

In addition, and as noted earlier, little research has been done to date on the potential application of RRI principles in a developing-country context. It is also important to note here that the principle of inclusion in the EC's version of RRI applies only to the inclusion of the views of *EC citizens* on future R&D imperatives. While that might help deliver R&D that is more aligned with societal challenges in Europe, perhaps even prioritizing work on climate change and green energy, by failing to include citizens of developing nations, the EC is unlikely to prioritize R&D investment in technology that is relevant to poverty in the South.

Finally, although RRI is potentially a useful part of the new toolkit for guiding technology innovation, it is not the only tool that will be needed. Even embedded in some form of regulation or standards, while it might enforce a degree of transparency and social inclusion in setting the direction of R&D and determine more appropriate areas for societal grand challenges, by itself it will not bring additional funding, particularly private-sector capital, to bear. Further, the issues identified in the last three chapters – innovation systems, intellectual property rights, and the role of the state – will still have to be addressed.

Innovation for whom? Inclusive innovation and the voices of the marginalized

Converging views from business and development sectors

Building on the concept of national innovation systems, a body of research under the label of 'inclusive innovation' has emerged in the academic literature, largely since 2011 (Heeks et al., 2013). The research has generally focused on developing countries and how innovation (and innovation systems) can be made more relevant to the needs of marginalized groups, which could include women, young people, and people with disabilities, among others, but which, in the literature to date, generally refers to 'the poor'. The focus here is on the question, 'Innovation for whom?'

The growth in interest in inclusive innovation has come from two directions. Firstly, from the management literature and the dawning realization by business of the possibilities offered by the 4 billion people living in households with less than $1,500 income a year (see, for example, Prahalad and Hart, 2002). This is mostly a top-down, company-led perspective, shaped by management studies thinking, of what businesses could do to shape their offers to meet the needs of poorer sectors of society. The second direction of research, however, has been more concerned with the perspectives of low-income communities themselves and how they might be actors in the innovation process, and is informed more by development studies and livelihoods approaches. A wide range of terms have been used in this second approach to describe how communities and small enterprises engage in innovation processes, including: frugal, indigenous, pro-poor, inclusive, local, grassroots, and informal innovation (Foster and Heeks, 2014). The term 'inclusive' can mean a range of different things which, together, can be thought of as a ladder of increasing levels of engagement of marginalized populations in the innovation process (see the classification shown in Box 11.1).

Inclusive innovation approaches

Examples of inclusive innovation in the literature are limited. Fressoli and colleagues (2014) provide three instances of inclusive (or grassroots) innovation from Brazil and India:

1. The Honey Bee Network (HBN) was founded in 1989 by a group of scientists and farmers in India to capture and raise awareness of indigenous agricultural practices and knowledge against the backdrop of an agricultural innovation narrative focused on Green Revolution

Box 11.1 Different levels of 'inclusivity' of innovation

Level 1/Intention: An innovation is inclusive if the intention of that innovation is to address the needs or wants or problems of the excluded group.

Level 2/Consumption: An innovation is inclusive if it is adopted and used by the excluded group. This requires the innovation to be developed into goods or services that can be accessed and afforded by the excluded group, which in turn has the motivation and capabilities to absorb the innovation.

Level 3/Impact: An innovation is inclusive if it has a positive impact on the livelihoods of the excluded group (which could mean improved productivity, wellbeing, or the creation of livelihood assets or capabilities).

Level 4/Process: An innovation is inclusive if the excluded group (or at least members of the group) are involved in some or all of the processes of development of the innovation (for example, invention, design, development, production, distribution).

Level 5/Structure: An innovation is inclusive if it is created within a structure that is itself inclusive. The argument here is that inclusive processes may be temporary or shallow in what they achieve. Deep inclusion requires that the underlying institutions and relations that make up an innovation system are inclusive. This might require either significant structural reform of existing innovation systems, or the creation of alternative innovation systems.

Level 6/Post-structure: An innovation is inclusive if it is created within a frame of knowledge and discourse that is itself inclusive. (Some) post-structuralists would argue that our underlying frames of knowledge – even our very language – are the foundations of power which determine societal outcomes. Only if the framings of key actors involved in the innovation allow for inclusion of the excluded can an innovation be truly inclusive.

Source: Heeks et al., 2013

technologies. The HBN is an informal network that also helps individuals explore the commercial potential of traditional products and processes with the protection of intellectual property rights, seed funding, and incubation services. It places weight on recognizing the rights of traditional knowledge-holders and ensuring that a fair and reasonable share of any proceeds accruing from the value addition of local traditional knowledge and innovation goes back to the knowledge-holders (HBN, 2015).

2. The Social Technologies Network (STN) in Brazil, which ran from the early 2000s to 2012, had a range of participants including academics, unions, government representatives, funding agencies, non-government organizations, and community groups. As such it was more of a mixture of mainstream large-scale technology players and local innovators than the HBN. Its principal aim was not that local communities had to be innovators themselves, but that technology developers needed to make sure poor communities were fully included in the innovation process and in adopting and benefiting from the arising technology. Two examples of technology programmes that were taken to scale through STN were the PAIS agroecology programme and the million cisterns programme. PAIS involved low-cost technology aimed at farms below 2 hectares in size and promoted local knowledge around agroecological

approaches to food production over the use of external inputs and pesticides. Farmers received a kit that included 'components for a water irrigation system, wire fences, seed, small plants and even hens, along with a user's manual and a training course' (Fressoli and Dias, 2014). PAIS was selected by the STN for application across 12 states. The million cisterns programme was an attempt to provide year-round access to water supplies in a large semi-arid region in northeast Brazil with a population of around 25 million people through the construction of a huge number of rainwater-harvesting cisterns or tanks that could store water throughout the dry season for domestic use.

3. The People's Science Movement (PSM), again from India, was started in the 1980s and focused on the potential for upgrading traditional knowledge and practice through the application of 'modern' science. Instead of concentrating only on technology development, the PSM's approach aimed to enable artisans and workers as 'carriers of technology' to organize themselves and acquire capabilities to upgrade their own technological skills and capacity. The grassroots approach to innovation included participatory technology development, pro-poor business models, and technical capability development in a systems approach to innovation. Like the STN, PSM was focused on concrete improvements for marginalized people in India and on the empowering of communities and artisans to tackle deeper structural change in innovation systems.

The introduction of the mobile phone to bottom-of-the-pyramid markets in Kenya is a fourth example of inclusive innovation referred to in the literature (see Foster and Heeks, 2013, 2014). Here the innovation is less about the technology itself and more about its means of dissemination to the poor, and particularly the role played by small enterprises, and novel uses of that technology to adapt it to low-income markets (ranging from the technique of 'beeping' – calling a contact at no cost by simply registering a missed call, through to the adoption of mobile money services).

Interestingly, a seemingly obvious example of inclusive innovation, Farmer Field Schools, doesn't seem to get any attention in the inclusive innovation literature. The Farmer Field School approach brings farmers together to learn for themselves through participation in their own on-farm trials and research, and using their own observations to learn about crop production problems and to develop ways to deal with them. The approach was first tried in Southeast Asia to improve pest control but was adapted for use in Africa by the Food and Agriculture Organization to work with farmers on a variety of problems, including improvements in soil productivity, conservation agriculture, soil and water conservation, and improved irrigation (FAO, 2015).

Lessons so far

The concept of inclusive innovation remains quite a wide one, as can be seen from the examples given and the ladder of inclusivity sketched out in Box 11.1.

Compared to the responsible innovation approach, both the development of analytical frameworks in the literature and the development of inclusive innovation principles for practical application seem to be at an early stage. What is consistent across research findings is an assertion that technology innovation will have a positive impact on the lives of poor and marginalized communities in the developing world only if their voices are heard at some point in the process.

Innovation by whom? Learning from the open-source movement

Open-source innovation versus open innovation

One alternative to intellectual property rights (IPR) as an approach for driving and managing innovation processes is modelled by the open-source movement. It is important to distinguish here the difference between 'open innovation' and 'open-source innovation' as the terms are sometimes confused.

The process of open innovation was named and first described by Henry Chesbrough, who suggested that companies traditionally favoured closed innovation processes where innovation was managed internally within their own R&D departments, using their own employees to develop new products and take them to market, while preserving the confidentiality felt necessary to secure IPR and gain commercial advantage. Chesbrough argued the opposite: that open innovation approaches, where external partners perform part of the innovation process, were more effective because no one company was likely to have a monopoly on the smartest people (Chesbrough, 2003). According to Chesbrough:

> Open innovation suggests that valuable ideas can come from inside or outside the company and can go to market from inside or outside the company as well. This approach places external ideas and external paths to market on the same level of importance as that reserved for internal ideas and paths to market in the earlier era. (Chesbrough, 2006: 2)

Although some argue that Chesbrough set up a false dichotomy between closed and open systems of innovation and that many companies had historically recognized the value of external knowledge (Trott and Hartmann, 2009), the 'external path to markets' for the intellectual property element of the model does distinguish the open innovation approach from previous thinking (and also distinguishes it from open-source innovation).

The business model for open innovation is not just about sourcing ideas from outside as well as in, but also the notion that the company might not always be the best vehicle to take its own innovations to market. The open innovation approach therefore puts significant emphasis on the development of intellectual property markets that allow firms to sell and export the knowledge and technologies they have developed themselves. This does stand in sharp contrast to traditional theories which consider knowledge and innovation as a core activity that should never be shared or sold (Pénin et al., 2011). But this continued reliance on patents and IPR as a key driver of behaviour means that open innovation approaches have limited potential to drive innovation in areas where there are weak market signals, such as the development of drugs for diseases of the developing world or agroecological technologies for food production.

The open-source approach, in contrast, does not rely on the trading of IPR as the key driver of innovation, which makes the process more interesting as a possible alternative to the problems thrown up by patents, described in Chapter 9. The term 'open-source' refers to, 'something that can be modified and shared because its design is publicly accessible'. As an approach its response to the question 'innovation by whom?' is 'for everyone and anyone' (Opensource.com, 2015)!

The origins of the term 'open-source' lie in the context of computer software development, referring to software with source code (the original code that programmers write in) available and accessible for anyone to use or alter, unlike proprietary software which kept this secret. The Linux operating system is probably one of the best known open-source products, created by Linus Torvalds but supported, 'debugged' and further developed by a whole community of volunteer programmers. In a famous paper, 'The cathedral and the bazaar', Eric Raymond described how Torvalds's approach differed from the traditional view that,

> the most important software (operating systems and really large tools like the Emacs programming editor) needed to be built like cathedrals, carefully crafted by individual wizards or small bands of mages working in splendid isolation, with no beta to be released before its time. (Raymond, 2000: 2)

Instead, Torvalds's style of development involved releasing beta versions of the software 'early and often', and delegating

> everything you can ... [and being] open to the point of promiscuity ... No quiet, reverent cathedral-building here – rather, the Linux community seemed to resemble a great babbling bazaar of differing agendas and approaches (aptly symbolized by the Linux archive sites, who'd take submissions from anyone). (Raymond, 2000: 2)

Essentially, Torvalds crowd-sourced the debugging and development of Linux, leading Raymond to coin Linus's Law: 'given enough eyeballs, all bugs are shallow' (Raymond, 2000: 9).

Open-source software is in widespread use today. Much of the internet is built on open-source software such as the Linux system and the Apache web server application (which has been the most popular web server on the internet for the past 20 years). Other popular open-source software includes the GIMP image-processing system, Open Office, the Android system for mobile phones, and the Google browser. Open-source software is not always free to the user and programmers can charge money for the open-source software they create or to which they contribute. However, open-source licences require programmers to release their source code when they sell software to others to allow further development and manipulation by third parties. For this reason, many open-source software programmers find it more profitable to charge users money for helping to install, use, or troubleshoot than to charge for the software itself.

Open-source software and licences

IPR remains an important issue with open-source software and licences are still a common feature for users to abide by. But here the purpose of the licence is not to protect the right of the individual or company to own and profit from intellectual property, but to *prevent* the private enclosure of that property and to keep it firmly in the public domain. Indeed, open-source software could not have flourished without the legal innovation represented by GNU General Public License, developed in the 1980s by Richard Stallman, a programmer from MIT (Kapczynski et al., 2005). The Open Source Initiative lists 10 defining features of open-source licensing that not only encourage experimentation and use of open-source products and code, but also place a legal responsibility on the user to maintain the rights and abilities of others to do the same (see Box 11.2).

Box 11.2 Open-source definition: distribution terms open-source software must abide by

1. **Free redistribution:** The licence shall not restrict any party from selling or giving away the software as a component of an aggregate software distribution containing programs from several different sources. The licence shall not require a royalty or other fee for such sale.
2. **Source code:** The program must include source code, and must allow distribution in source code as well as compiled form.
3. **Derived works:** The licence must allow modifications and derived works, and must allow them to be distributed under the same terms as the licence of the original software.
4. **Integrity of the author's source code:** The licence must explicitly permit distribution of software built from modified source code. The licence may require derived works to carry a different name or version number from the original software.
5. **No discrimination against persons or groups:** The licence must not discriminate against any person or group of persons.
6. **No discrimination against fields of endeavour:** The licence must not restrict anyone from making use of the program in a specific field of endeavour. For example, it may not restrict the program from being used in a business, or from being used for genetic research.
7. **Distribution of licence:** The rights attached to the program must apply to all to whom the program is redistributed without the need for execution of an additional licence by those parties.
8. **Licence must not be specific to a product:** The rights attached to the program must not depend on the program's being part of a particular software distribution.
9. **Licence must not restrict other software:** The licence must not place restrictions on other software that is distributed along with the licensed software.
10. **Licence must be technology-neutral:** No provision of the license may be predicated on any individual technology or style of interface.

Source: Open Source Initiative, 2007

In recent years the concept of open-source has started to be applied beyond the software development sector and today the term is often used to designate a set of values. For example, according to the online publication Opensource.com (2015), 'open-source projects, products, or initiatives are those that embrace and celebrate open exchange, collaborative participation, rapid prototyping,[6] transparency, meritocracy, and community development'. Three examples of wider applications are provided in the next section.

3D printing

3D printers work by building up layers of polymer or resin to form a three-dimensional object. Driven by computer-aided design (CAD) software, the printer head moves in two dimensions to lay down a film of material on a platform. The platform is then lowered slightly to allow the head to repeat the process, creating the next layer, and so on. Complex objects can be produced in this manner and, although most of the cheaper printers available are restricted to printing with plastics to create objects no bigger than a small shoe box, some commercially used printers can print much larger objects and also use metal alloys as opposed to plastics as the 'ink'.

3D printing is supported by open-source software, such as Blender, for creating 3D CAD files and by a wide range of enthusiasts and communities developing, sharing, and improving CAD software files to print specific objects. People have not been slow to see the potential for open-source sharing and improving of 3D-printing files over the internet as a means of expanding access to appropriate technologies in the developing world (Pearce et al., 2010). A competition to develop 3D printing applications for the developing world held in London in 2012 attracted a wide range of entries, including applications to produce specially designed 3D-printed shoes for individuals suffering from foot deformities due to jigger fly infestation, printed parts for solar lights, a set of cheap and rapidly manufactured soft tissue prostheses, and parts for water supply and sanitation systems printed from recycled plastic (techfortrade, 2012).

Perhaps the most well-known open-source project in 3D printing at the moment is RepRap, short for Replicating Rapid Prototype. Founded by Adrian Bowyer, an engineer and former academic at Bath University in the UK, RepRap is an attempt to make the world's first self-replicating machine – a 3D printer that can print a copy of itself (Jones et al., 2011). At the time of writing RepRap is able to print about 50 per cent of its parts from plastic, with efforts now focused on being able to replicate some of the electronics, too. The RepRap team, supported by a community of users and developers, continues to develop and to give away the designs for this cheap 3D printer (material costs are about €350). The aim is to make the technology ultimately accessible to small communities in the developing world as well as individuals in the developed world. Following the principles of the free software movement, the plans, instructions, and software for the RepRap machine are distributed at no cost to everyone under an open-source licence so, 'if you have a RepRap machine, you can use it to make another and give that one to a friend' (RepRap, 2014).

Although 3D-printing technology is still in its infancy, the possibility it may offer to put manufacturing capabilities into the hands of individuals anywhere in the world, supported by access through the internet to a huge and ever-expanding library of intellectual property in the form of open-source 3D-printer files, has led commentators such as the American economist Jeremy Rifkin to speculate about a global 'third industrial revolution', where the means of production shifts away from big corporations and back to communities and individuals (Rifkin, 2011).

Development of seeds

Today, it is estimated that 56 per cent of the global proprietary seed market is controlled by just four transnational companies (Howard, 2009). This consolidation of ownership is associated with negative impacts on opportunities for sustainable agriculture, including reductions in the number of seed lines and a decline in the practice of seed saving,[4] as the influence of these companies over farmer behaviour increases. Concerns are also mounting over new interpretations of patent laws that allow greater corporate control over not only genetically engineered plants, but also over seed and plant material created through traditional breeding processes (for example, the patent granted by the European Patent Office in 2013 to Syngenta for pepper varieties produced through conventional breeding (Saez, 2013)) or even plants discovered in the wild (such as the patent Monsanto obtained in 2014, again from the European Patent Office, granting it a monopoly on the future usage of hundreds of natural DNA sequence variations in the conventional breeding of cultivated and exotic species of soybeans (No patents on seeds, 2014)).

The Open Source Seed Initiative (OSSI) was formed by a group of plant breeders, farmers, seed companies, and sustainability advocates in response to this growing use of patents by a handful of powerful companies to increase their control over global seed markets. Noting that patented seeds cannot be saved, replanted, shared by farmers, or used by universities and breeders to create new varieties, and inspired by the open-source software movement, 'OSSI was created to free the seed – to make sure that the genes in at least some seed can never be locked away from use by intellectual property rights' (Open Source Seed Initiative, 2015). Although falling short of a full legal licence, OSSI has created an open-source 'pledge' that farmers and breeders can use in order to ensure the availability of their seed lines to other breeders now and in the future. The pledge states:

> You have the freedom to use these OSSI-pledged seeds in any way you choose. In return, you pledge not to restrict others' use of these seeds or their derivatives by patents or other means, and to include this pledge with any transfer of these seeds or their derivatives. (Open Source Seed Initiative, 2015)

Medical research

A Yale Law School paper on open licensing in the health sector noted a rise in 'commons-based' forms of production and coordination 'that rely on a mechanism other than proprietary exclusion and that treat all actors symmetrically vis-a-vis the resource in question' (Kapczynski et al., 2005: 1068). The Human Genome Project is one such notable example which was publicly funded and released its data without claiming any patent rights. The HapMap project, a catalogue of common genetic variants that occur in human beings, is also commons-based. The HapMap makes its data available for free on the internet and takes the additional step of 'creating a click-wrap license[5] to prevent those accessing its data from combining it with their own data and patenting the results' (Kapczynski et al., 2005: 1071).

Indeed, given the historical influence of global pharmaceutical corporations' proprietary practices, there is a surprising amount of open-source activity emerging in the pharmaceutical sector as academics, civil-society organizations, foundations, governments, and even some pharmaceutical companies try to find a way around the market failures related to neglected diseases. Three examples of open-source activity in this area are:

The Structural Genomics Consortium (SGC). The SGC is a not-for-profit, public–private partnership with a mission 'to determine 3D structures of human proteins of biomedical importance and proteins from human parasites that represent potential drug targets'. The SGC is a consortium of six universities[6] which are home to some 200 scientists and support staff. SGC is funded by 13 separate organizations that include major drug companies, such as GlaxoSmithKline and Pfizer; foundations, such as the Wellcome Trust; and the Ontario Ministry of Economic Development and Innovation. The SGC accelerates research by making all its research output available to the scientific community with no patents or IPR restrictions attached, and by 'creating an open collaborative network of scientists in hundreds of universities around the world and in nine global pharmaceutical companies'. Up to 2011, the SGC had released for public use the structures of over 1,200 proteins with implications for the development of new therapies for cancer, diabetes, obesity, and psychiatric disorders (SGC, 2015).

The Open Source Malaria (OSM) project. OSM works on improving the properties of compounds that are in the public domain and that are known to have the potential to kill the malaria parasite in cells, in order to discover a compound that can enter early-stage clinical trials. The project operates on open-source principles in that everything is open. Funded by the Australian government, the project is available to anyone who wants to take part and there is a list of tasks that need doing on its website ranging from administrative and writing tasks to chemical synthesizing. Currently active contributors range from academic researchers at universities in Australia, Canada, India, Spain, Sweden,

Switzerland, and the UK, to private companies and even students from an Australian grammar school (OSM, 2015).

The Open Source Drug Discovery initiative (OSDD). OSDD is an initiative aimed at affordable health-care launched by the government of India and led by the Council of Scientific and Industrial Research, one of the largest publicly funded research organizations in the world. Tuberculosis was OSDD's first target for drug discovery, due to its high incidence and mortality in India and other developing countries, and because no advances had been made to address the problems of the existing treatment regime. The existing treatment involves a mixture of up to four different antibiotics over at least six months. The length of treatment and side effects from the drugs pose huge problems for tuberculosis patients and for global efforts to tackle the disease (TB alert, 2015). OSDD has added malaria as a second research area. Made up of a community of more than 7,900 participants from 130 countries, including students and research scientists from academia and industry, OSDD provides an open innovation research platform for both computational and experimental technologies. All projects and research results are reported publicly via a web-based platform that enables independent researchers to freely share their work and collaborate over the internet. The approach used is to conduct early-stage research in a very open and highly collaborative environment via the web-based platform. As promising compounds are identified to move to the development stage, OSDD narrows collaboration down to contracted private or public-sector pharmaceutical research organizations, mostly in India, to keep costs down and build national capacity. For final delivery of drugs to market, OSDD intends to rely on the generic drug industry, making sure that all drugs emerging from its initiative will be made available in generic form, without any intellectual property restrictions (OSDD, 2015).

Motivation in open-source approaches to innovation

As these examples have shown, there is a growing number of instances of technology innovation being achieved through open and collaborative mechanisms, a fact that runs counter to the conventional wisdom that knowledge needs to be captured, enclosed, and privatized through patents to provide the necessary commercial stimulus to advance science. But what motivates individuals and institutions to collaborate in this manner when the proprietorial incentive is removed?

Studies in the software industry (see ; Bonaccorsi and Rossi, 2003; Andersen-Gott, 2011; Ye and Kishida, 2013) suggest that individual programmers are motivated to contribute to open-source projects because of philosophical commitments to the concept and because of the potential for personal development – notably by improving their skills through learning by doing, building up a personal curriculum vitae (résumé), and advertising their skills and prowess in a forum where prospective employers may be present. In turn, corporates who engage in crowd-sourcing most often cite cost (open-source development as a way to allow small enterprises to afford innovation) and quality (the contributions and feedback from the open-source community help to remove bugs and improve the reliability and quality of the software) as reasons for participating.

The motivations of those involved in the Open Source Seed Initiative may be more altruistic and related to broader commitments to the maintenance of biodiversity, but even here there are business-related drivers. The initiative is a conscious attempt to reclaim and maintain some space for independent small seed producers in a market that is rapidly being enclosed in patent thickets created by the big transnational seed companies.

In the case of the Kaggle platform mentioned in Chapter 9, again there seem to be multiple motivations. Predictive data analysts need sources of raw data against which to test their new models and often spend a lot of time searching for and then cleaning up new datasets to use for their research. The datasets and predictive problems from commercial clients provided by Kaggle offer opportunities for analysts to spend more time on research and less time hunting new datasets. Layered on top of this are opportunities for financial rewards if you provide the winning solution in a competition and, again, opportunities to display your skills in front of prospective employers. In return, companies get access to a much wider set of possible solutions than they would be able to generate internally, and, sometimes, exposure to new approaches they were not even aware of. For example, the pharmaceutical company Merck worked with Kaggle to streamline its drug discovery process, which often involves testing hundreds of thousands of compounds for different diseases. Merck set up a competition based on data on chemical compounds it had previously tested, and challenged participants to develop a predictive model that could identify which compounds had the greatest potential for further testing. According to the *Harvard Business Review,*

> the contest attracted 238 teams that submitted well over 2,500 proposals. The winning solution came from computer scientists (not professionals in the life sciences) employing machine-learning approaches previously unknown to Merck. (Boudreau and Lakhani, 2013)

Harnessing the energy of open source

Surprisingly, it is the health sector that provides some of the most compelling examples outside the software industry of why open-source approaches to technology innovation may have something genuine to offer, whether it is to developing-country governments trying to find quicker ways to deliver new generic and cheap drugs for neglected diseases, or even big pharmaceutical corporations trying to find a lower cost route to market. Perhaps because the pharmaceutical sector is the most proprietorial of proprietary businesses, it also displays many of the strongest rationales for a change to open-source. The enormous cost of clinical trials and the low success rate of promising drugs once they are entered into that process translates into an incredibly inefficient use of resources when done in secret. Multiple companies can be working on the same problem at the same time, investing in repeating the same research and, in some cases, continuing a line of investigation for many years without access to knowledge that others have already abandoned it for good reason.

But, as the examples given here show, open-source innovation is not restricted to a single sector. The appeal of the approach is spreading. The growth in free-to-access massive online open courses, or MOOCs, led by elite US academic institutions that see opportunities to enhance brand, experiment with teaching approaches, boost recruitment, and innovate new business models, is another example (BIS, 2013). This aligns with other movements towards the application of open-source principles in the education sector, such as the Open Education Consortium. This broader adoption of open-source-driven innovation suggests that it has an important part to play in a more just approach to the governance of science-based technology development and deployment, which has the potential to reach places and service needs that market-driven innovation has so far failed to reach, and to improve the efficiency and effectiveness of technology development processes even where strong market pulls already exist.

Notes

1. Sometimes referred to as 'responsible innovation' or RI.
2. For example, neither the general ideas of RRI nor any specific reference to it feature in documentation published so far on the SDG Technology Facility Mechanism, the role and activities of the UN interagency Task Team on Science, Technology and Innovation for the SDGs (UN Technology Facilitation Mechanism, 2015), or the original mandate of the mechanism (UN General Assembly, 2015). Likewise, the concept is absent from the most relevant documentation of the UNFCCC Technology Mechanism, such as the TEC's briefing note on innovation systems in developing countries (TEC UNFCCC, 2015).
3. Rapid prototyping refers to the process of being prepared to try things out quickly, accepting that this may often lead to rapid failure, but using that process repeatedly to learn from those failures to ultimately produce a better outcome.
4. Seed saving is the practice of holding back part of a harvest to provide a source of seed for planting the following year.
5. A click-wrap licence is commonly found in software, where the user has to click on a button to agree to conditions of a licence before gaining access to the software itself.
6. Universities of Toronto (Canada), Oxford (UK), Frankfurt (Germany), Campinas (Brazil), Karolinska (Sweden), and North Carolina (US).

References

Andersen-Gott, M. (2011) 'Why do commercial companies contribute to open source software?' [online], November 8, <http://www.sciencedirect.com/science/article/pii/S026840121100123X> [accessed 27 November 2015].

Bonaccorsi, A. and Rossi, C. (2003) 'Comparing motivations of individual programmers and firms to take part in the Open Source movement. From community to business' [online], November 19, <http://papers.ssrn.com/sol3/papers.cfm?abstract_id=460861> [accessed 27 November 2015].

Boudreau, K.J. and Lakhani, K.R. (2013) 'Using the crowd as an innovation partner', *Harvard Business Review,* April <https://hbr.org/2013/04/using-the-crowd-as-an-innovation-partner/> [accessed 16 March 2016].

Chesbrough, H. (2003) *Open Innovation: The New Imperative for Creating and Profiting from Technology*, Boston, MA: Harvard Business School Press.

Chesbrough, H. (2006) *Open Innovation – Researching a New Paradigm,* Oxford: Oxford University Press.

Department of Business Innovation and Skills (2013) *The Maturing of the MOOC: Literature Review of Massive Open Online Courses and Other Forms of Online Distance Learning,* London: Department for Business, Innovation & Skills, UK Government.

European Commission (2012) *Responsible Research and Innovation: Europe's Ability to Respond to Societal Challenges,* Brussels: European Commission.

European Commission (2013) 'Fact Sheet: Science with and for society' [online], *Europa* <https://ec.europa.eu/programmes/horizon2020/sites/horizon2020/files/FactSheet_Science_with_and_for_Society.pdf> [accessed 18 November 2015].

European Commission (2015a) *Research & Innovation: Participant portal H2020 online manual* [online], *Europa* <http://ec.europa.eu/research/participants/docs/h2020-funding-guide/index_en.htm> [accessed 20 November 2015].

European Commission (2015b) *Science with and for Society* [online]. Horizon 2020 The EU Framework Programme for Research and Innovation: <https://ec.europa.eu/programmes/horizon2020/en/h2020-section/science-and-society> [Accessed 18 November 2015].

FAO (2015) 'Land resources: Farmer Field School', [online], *FAO* <http://www.fao.org/nr/land/sustainable-land-management/farmer-field-school/en/> [accessed 30 November 2015].

Fisher, E., Mahajan, R., and Mitchum, C. (2006) 'Midstream modulation of technology governance from within', *Bulletin of Science, Technology and Society,* 26(6): 485–96 <http://dx.doi.org/10.1177/0270467606295402>.

Foster, C. and Heeks, R. (2013) 'Conceptualising inclusive innovation: modifying systems of innovation frameworks to understand diffusion of new technology to low-income consumers', *European Journal of Development Research,* 25: 333–55 <http://dx.doi.org/10.1057/ejdr.2013.7>.

Foster, C. and Heeks, R. (2014) 'Nurturing user-producer interaction: innovation flows in a low income mobile phone market', *Innovation and Development,* 4: 221–37 <http://dx.doi.org/10.1080/2157930X.2014.921353>.

Fressoli, M. and Dias, R. (2014) *The Social Technology Network: A Hybrid Experiment in Grassroots Innovation,* Brighton: STEPS.

Fressoli, M., Arond, E., Abrol, D., Smith, A., Ely, A., and Dias, R. (2014) 'When grassroots innovation movements encounter mainstream institutions: implications for models of inclusive innovation', *Innovation and Development,* 4(2): 277–92 <http://dx.doi.org/10.1080/2157930X.2014.921354>.

Grunwald, A. (2004) 'Vision assessment as a new element of the FTA tool box', *EU–US Seminar: New Technology Foresight, Forecasting & Assessment Methods – Seville 13-14 May 2004* (pp. 53–67), Seville: European Commission.

HBN. (2015) 'About us', *Honey Bee Network*: <http://www.sristi.org/hbnew/aboutus.php> [accessed 30 November 2015].

Heeks, R., Amalia, M., Kintu, R., and Shah, N. (2013) *Inclusive Innovation: Definition, Conceptualisation and Future Research Priorities*. Manchester: SEED, University of Manchester.

Howard, P.P. (2009) 'Visualizing consolidation in the global seed industry: 1996–2008', *Sustainability,* 1(4): 1266–87 <http://dx.doi.org/10.3390/su1041266>.

Jones, R., Haufe, P., Sells, E., Iravani, P., Olliver, V., Palmer, C., and Bowyer, A. (2011) RepRap – the replicating rapid prototyper', *Robotica,* 29(1): 177–91 <http://dx.doi.org/10.1017/S026357471000069X>.

Kapczynski, A., Chaifetz, S., Katz, Z., and Benkler, Y. (2005) 'Addressing global health inequalities: an open licensing approach for university innovations', *Berkley Technical Law Journal,* 20(2): 1031–88 <http://www.jstor.org/stable/24116640>.

No Patents on Seeds (2014) 'Opposition against Monsanto patent EP2134870 on the selection of soybeans', *No Patents on Seeds,* <https://no-patents-on-seeds.org/en/information/background/opposition-against-monsanto-patent-ep2134870-selection-soybeans> [accessed 26 November 2015].

Open Source Initiative (2007) 'Open source definition', 2 March, *Open Source Initiative,* <https://opensource.org/osd> [accessed 24 November 2015].

Open Source Seed Initiative (2015) 'The open source seed initiative', *Open Source Seed Initiative,* <http://osseeds.org/> [accessed 26 November 2015].

Opensource.com (2015) 'What is open source?', *Opensource.com,* <http://opensource.com/resources/what-open-source> [accessed 23 November 2015].

OSDD (2015) 'Collaborative innovation platform', *Open Source Drug Discovery,* <http://www.osdd.net/about-us/how-osdd-works> [accessed 27 November 2015].

OSM (2015) 'Open Source Malaria landing page', *Open Source Malaria,* <http://opensourcemalaria.org/#> [accessed 26 November 2015].

Owen, R., Macnaghten, P., and Stilgoe, J. (2012) 'Responsible research and innovation: from science in society to science for society, with society', *Science and Public Policy,* 39: 751–60 <http://dx.doi.org/10.1093/scipol/scs093>.

Pearce, J., et al. (2010) '3-D printing of open source appropriate technologies for self-directed sustainable development', *Journal of Sustainable Development,* 3(4): 17–29.

Pénin, J., Hussler, C., and Burger-Helmchen, T. (2011) 'New shapes and new stakes: a portrait of open innovation as a promising phenomenon', *Journal of Innovation Economics and Management,* 2011/1(7): 11–29.

Prahalad, C.K. and Hart, S.L. (2002) 'The fortune at the bottom of the pyramid', *Strategy & Business*, 26 <http://www.strategy-business.com/article/11518?gko=9a4ba> [accessed 16 March 2016].

Raymond, E.S. (2000) 'The cathedral and the bazaar', <http://www.catb.org/~esr/writings/cathedral-bazaar/cathedral-bazaar/> [accessed 24 November 2015].

RepRap (2014) 'About', <http://reprap.org/wiki/About> [accessed 26 November 2015].

Rifkin, J. (2011) *The Third Industrial Revolution*, New York: Palgrave Macmillan.

Saez, C. (2013) 'EPO still granting patents on conventional vegetables: "just following the rules"', 24 May, *Intellectual Property Watch*, <http://www.ip-watch.org/2013/05/24/epo-still-granting-patents-on-conventional-vegetables-just-following-rules/> [accessed 26 November 2015].

SGC (2015) 'Structural Genomics Consortium', <http://www.thesgc.org/> [accessed 27 November 2015].

Stilgoe, J., Owen, T., and Macnaghten, P. (2013) 'Developing a framework for responsible innovation', *Research Policy*, 42(9): 1568–80.

Sutcliffe, H. (2015) *Principles for Responsible Innovation: Building Trust and Trustworthiness in Business Innovation*, London: MATTER.

TB alert (2015) 'Treatment', *TB alert*, <http://www.tbalert.org/about-tb/what-is-tb/treatment/> [accessed 27 November 2015].

TEC UNFCCC (2015) 'Strengthening national innovation systems to enhance action on climate change', TEC Brief #7, Bonn: Technology Executive Committee, United Nations Framework Convention on Climate Change.

techfortrade (2012) '3d4d challenge finalists announced', *techfortrade*, <http://techfortrade.org/news/3d4d-challenge-finalists-announced/> [accessed 26 November 2015].

Trott, P. and Hartmann, D. (2009) 'Why "open innovation" is old wine in new bottles', *International Journal of Innovation Management*, 13(4): 715–36.

UN General Assembly (2015) 'Transforming our world: the 2030 agenda for sustainable development', clause 70, *UN Sustainable Development Knowledge Platform*, <https://sustainabledevelopment.un.org/post2015/transformingourworld> [accessed 25 January 2016].

UN Technology Facilitation Mechanism (2015) 'Terms of reference for the UN Interagency Task Team on Science, Technology and Innovation for the Sustainable Development Goals', *Technology Facilitation Mechanism*, <https://sustainabledevelopment.un.org/content/documents/8569TOR%20IATT%2026%20Oct%202015rev.pdf> [25 January 2016].

Ye, Y. and Kishida, K. (2013) 'Toward an understanding of the motivation of open source software developers', *Proceedings of 2003 International Conference on Software Engineering, May 3–10*, Portland, Oregon.

CHAPTER 12

Making technology innovation work for people and planet: the need to retool

Part II of this book has explored the drivers of technology innovation and shown that, today, there is a disconnect between the centre of gravity of the global technological innovation effort and the great challenges of ending poverty and finding a path to environmental sustainability. Instead of keeping humankind within that safe, inclusive, and sustainable space for development which lies between Raworth's minimum social foundation and Rockström's planetary boundaries, efforts to innovate often simply serve to further entrench inequalities and exacerbate the damage already being done to the ecosystems on which we rely for our very existence. The pattern of grave technology injustice that was revealed in Part I, where lack of access to technologies and the negative impacts of technology use were explored, are repeated and further reinforced when technological innovation processes are considered.

National innovation systems approaches reveal the wide range of institutions and relationships that are necessary to support successful technological innovation at a national level and, therefore, the range of policies and incentives that need to be in place to optimize this ecology of actors. Yet, in practice, an over-reliance on just two instruments to drive innovative behaviour – intellectual property rights and private capital – is pushing technological innovation in the wrong direction. This has to change.

The key challenge of climate change is the relatively small window of opportunity to cap greenhouse gas emissions before irreversible global warming sets in of a scale that could have catastrophic consequences for humankind. If that problem alone were not big enough, the path to avoiding climate change is inextricably linked to alleviating poverty in the developing world. Greenhouse gas emissions will only be kept within necessary limits if a global agreement is reached on their management, which will require, in turn, credible assurances for the developing world that their citizens will not be penalized as a result. That means credible support mechanisms need to be provided to enable those countries to follow a clean development path that will lead to a reasonable standard of living for all their citizens, as quickly as possible. It cannot be emphasized enough how important technology innovation will be in this process. It is vital that the global innovation effort is facing in the right direction, solving problems rather than creating – particularly in the case of food, energy, and water systems – additional climate and poverty burdens.

It is just as difficult to overemphasize how radical a step-change is needed in the technology innovation effort, to not only refocus efforts onto the challenges of poverty and sustainability, but also to scale up these efforts so that they may fully address these problems in the limited time available. To achieve this, a new set of tools is required to help govern and direct technology innovation: a set of tools that aligns with the principles of Technology Justice. Firstly, a practical and useable reworking of the precautionary principle is needed that will help us better anticipate the risks associated with the development of new technologies. Beyond that, new tools are needed that will help society shape the very purpose and direction of research and development, ensuring its attention stays focused, laser-like, on the two great challenges. This will require, among other things:

- The development of a new suite of incentives to either supplement or replace the existing and highly problematic intellectual property rights (IPR) system, including a scaling up of such instruments as prizes, advanced market commitments, direct research grants, and patent pools;
- A repurposing of international trade agreements in favour of building technological capacity in the developing world, and a stiffer spine and sharper set of teeth for the new UN Technology Bank to facilitate the development of technical capabilities in the developing world, as opposed to simply maintaining the status quo of power with regard to IPR;
- A new public–private alliance to be fashioned, with clear recognition of the entrepreneurial role of the state as a risk-taker and backer of innovation in situations of high risk and uncertainty, and clear incentives for the private sector to engage in the sharing of both risk and reward.

Strong global institutions are necessary to identify the critical areas of need for innovation and to help prioritize and coordinate resource flows and research agendas. At the moment these do not exist. The nearest example we have is the WHO's Global Observatory on Health Research and Development, which is successfully identifying global health research priorities. But even this mechanism has yet to move to being allowed to coordinate the necessary resources to bring about solutions. As things stand, the other major global technology mechanisms – including those of the UN Framework Convention on Climate Change and the Technology Facilitation Mechanism and Technology Bank of the UN's Sustainable Development Goals process – seem to be more focused on acting as clearing houses for information on existing technologies, rather than driving and shaping innovation agendas for new ones.[1]

Building on the responsible research and innovation approach, new – or heavily remodelled – global institutions will have to answer the questions of innovation: For what? For whom? And by whom? Marginalized voices will have to be brought in by these institutions to help ensure global research and development agendas properly address issues of poverty and exclusion as well as environmental sustainability that otherwise do not garner sufficient attention. In addition, similar institutional arrangements are likely to be needed at a national level to support more effective development of, and to give direction to, national innovation systems. Inclusive and open-source approaches to innovation provide some ideas around how this could be managed at a national and global level. Open-source innovation approaches have the potential to involve a wide range of viewpoints and voices in the innovation process, and, as shown in the last chapter, to provide degrees of speed and efficiency that, in some cases, traditional market-based competitive (but secret and parallel) processes cannot.

Too much technology innovation is heading in the wrong direction. The outdated instrument of market forces primarily steering that direction is not fit for purpose given the urgency and nature of the task at hand. It is time to retool: to develop a set of institutions and governance tools that *are* powerful enough to accomplish the task.

Note

1. The Technology Facilitation Mechanism (TFM) under the SDGs has, in theory, both a United Nations interagency task team and a collaborative annual multistakeholder forum which are focused on 'science, technology and innovation for the sustainable development goals'. These two groups are supposed to manage seven work streams, including 'Mapping of STI initiatives, background research and reports' (work stream 5) and a 'UN capacity building programme on technology facilitation for SDGs' (work stream 6) (UN Sustainable Development Knowledge Platform, 2015a). Very little material is available on how the TFM will work, but early terms of reference for the interagency task team suggest it will have very limited capacity (less than 10 part-time staff to support an annual meeting of a wider stakeholder group) and a restricted focus primarily on improved co-ordination within the UN system on technology and some work to support an 'online platform' (UN Sustainable Development Knowledge Platform, 2015b).

References

UN Sustainable Development Knowledge Platform (2015a) 'Technology facilitation mechanism', <https://sustainabledevelopment.un.org/topics/technology/facilitationmechanism> [accessed 26 January 2016].

UN Sustainable Development Knowledge Platform (2015b) 'Terms of reference for the UN Interagency Task Team on Science, Technology and Innovation for the Sustainable Development Goals', <https://sustainabledevelopment.un.org/content/documents/8569TOR%20IATT%2026%20Oct%202015rev.pdf> [accessed 25 January 2016].

PART III: REBOOT

Building a different approach to the governance of technology

CHAPTER 13
Reimagining technology as if people and planet mattered

The need to reboot our relationship with technology

The burden of technology injustice

This is a critical point in history, a point at which momentous decisions have to be taken that will determine the very survival of human (and other) species. The rapidly increasing global population contains a very large number of people who remain a long way from achieving the most basic standard of living and whose consumption of material resources (food, energy, water, and so on) consequently needs to increase. At the same time, though, human activity and resource consumption are already breaching planetary boundaries, opening the possibility of irreversible and catastrophic global environmental change.

A safe, inclusive, and sustainable space for humanity to occupy needs to be found, between a universally accessed minimum social foundation and a safe planetary environmental ceiling. Choices made about how technology is developed, disseminated, and used will have huge consequences for humanity's ability to establish that safe space. When everyone has access to the technologies needed for a minimum social foundation but the use of technology is governed to ensure this happens within those safe planetary boundaries is when we will have reached Technology Justice (see Figure 13.1).

This book has explored many examples of technology *injustice*. It has considered the social foundation and shown examples from the water, energy, food, health, and communication sectors where the world still falls far short of universal access to the technologies needed for a minimum standard of living. It has looked at injustices in the use of technology today, ranging from unintended negative impacts (such as the link between biofuel use and food prices or between the misuse and efficacy of antibiotics) to ways in which our use of technology is breaching planetary boundaries (for example, fossil fuels and climate change or industrial agricultural technologies and biodiversity). Finally, it has looked at technology innovation and the challenges faced by humanity in making sure research and development actually leads to technologies that address the social foundation and planetary boundaries problems.

The terms 'justice' and 'injustice' deliberately denote, in a Rawlsian sense, justice as 'fairness'. The examples considered, whether they relate to failures in the processes of technology access, use, or innovation, are not simply unfortunate occurrences or natural outcomes of some meta-system that humanity has no control over. They are the consequences of decisions made by human beings, by us. We could choose to put a price on carbon that

http://dx.doi.org/10.3362/9781780449043.014

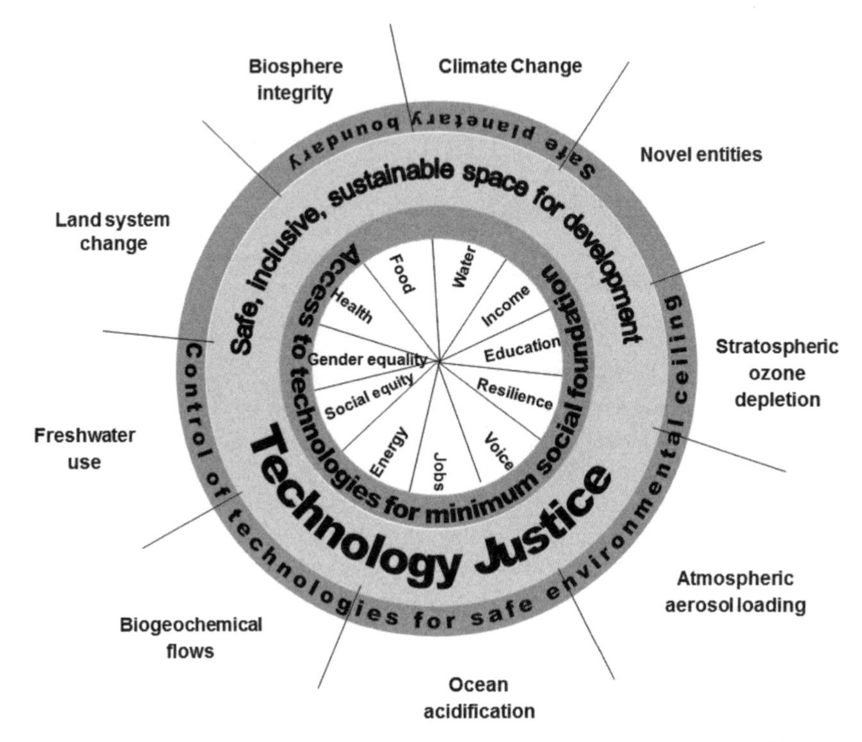

Figure 13.1 Technology Justice – a safe, inclusive, and sustainable space for humankind

would make fossil fuel technologies economically untenable and force the pace of development of key renewable technologies, such as storage. But we do not. We could choose to prioritize finance to extend vital services, such as energy, water, and sanitation, to those who have no access, rather than finance improvements to services for those who already have them. But we do not. We could choose to collaborate on research into neglected diseases by not only agreeing a global set of priority research areas, but by backing that with a set of pooled resources. But we do not.

This failure to make the right choices, to take responsibility for the governance of technology, has negative consequences for large numbers of people today, particularly the poor and marginalized in developing countries, who are the first to feel the impacts of the misuse of technology and the last to see the benefits of investments in research or dissemination. That more than 1 billion people still don't have access to the light bulb, a technology in common use for over a century, or that it is easier to raise money for research into male baldness than a vaccine for malaria, can only be described as monumental injustices. Moreover, the failure to make the right choices has huge consequences for future generations. At least three of the great technological leaps forward of the last 150 years – the technologies associated with fossil fuels, antibiotics, and the Green Revolution – have created unprecedented levels of environmental and social problems for present generations and left an unjust and toxic legacy for future generations to struggle with. In a very short space of time, the technologies that have transformed the living standards and lifespans of at least two-thirds of people on the planet already appear obsolete, while our slowness to recognize the scale of the technological challenge ahead risks the perpetration of an intergenerational injustice that could bring humanity to the cusp of disaster.

Time for a reboot

It is not the intention here to argue that issues of poverty and environmental sustainability can be addressed simply through the application of the correct forms of technology. Poverty and the associated lack of access to technology mirror broader social injustices and reflect the distribution of socio-economic power within and between different groups and nation states. While improving access to certain forms of technology can improve standards of living, poverty will not truly be addressed without tackling these imbalances in power and control over resources at all levels of society. Likewise, while a massive technological transformation is undoubtedly required to halt climate change, our ability to enact this is itself dependent on the success or otherwise of a huge political and social change project.

That said, both human wellbeing and the possibility of a sustainable future for life on this planet are tightly bound to the technological choices people make. The way humanity chooses to govern the development, dissemination, and use of technology is therefore critical. Moreover, technology choice has the potential to be subversive of existing power structures and should not therefore be dismissed entirely as an agent of socio-political change. A policy decision to support household solar photovoltaic electricity generation, whether in Germany or Kenya, is not simply a technology choice, it is also a choice to distribute ownership of the means of production (in this case energy generation) across a wider population rather than leaving it solely in the hands of a utility or private corporation. Similarly, support for research into agroecological techniques assumes a very different set of relationships between farmers and agricultural suppliers than research into conventional Green Revolution proprietary technologies.

Humanity has lost control of technology, or rather relinquished it to the vagaries of the market, assuming its 'invisible hand' will ensure the most efficient development and dissemination of technology that best meets people's needs – an assumption that is wrong. National innovation systems are scrutinized to see how a country's competitiveness can be improved in international science and technology markets instead of thinking about how to improve its capabilities to innovate in a socially and environmentally useful manner. People worry about managing the risks associated with the development of new technologies but not about whether the net impact of our science and technology effort is heading in the right direction. The world continues to be held hostage to an intellectual property regime dreamt up by Hollywood and the entertainment and software industries, despite evidence that it has no positive effect on innovation and may do more damage than good. And there is a constant underplaying of the entrepreneurial role government has to play in both innovation and dissemination of technologies where market signals are weak or non-existent. This is most notable in the areas where technology is urgently needed to deal with the two greatest challenges facing humanity – ending poverty and ensuring environmental sustainability.

While needing to avoid falling into the trap of technological determinism, humanity has to reboot its relationship with technology. This is not incremental change but a radical shift in the oversight and governance of innovation, and access to and use of technology. The lens of Technology Justice enables us to recognize that some choices are more likely to lead to that safe and equitable space for human development, whereas other choices are more likely to lead in the opposite direction. Responsibility needs to be taken for those decisions rather than hoping market mechanisms can make them by default and without intervention. That does not mean there is not a role for markets or the private sector – far from it. But it does require two things:

- Recognition that in the areas of most concern – poverty and the environment – markets alone will not provide the 'pull' mechanisms to deliver the right technologies to the right people. Those mechanisms and markets will have to be created through entrepreneurial actions of governments and supported, often for longer than conventional wisdom currently dictates, to create the conditions for private capital to flow and to deliver the change necessary;
- Recognition that in those areas where progress is most urgent, new forms of incentivizing innovation that make best use of all the physical and intellectual capabilities available need to be sought. This may point towards more collaborative and open-source approaches than have been the case in the past.

So what would a reboot of our relationship with technology look like? How would access, use, and innovation be approached differently if Technology Justice was the main goal and driver of this new relationship – if technology was approached as if people and planet truly mattered? The following section draws on the analysis of the earlier chapters to suggest what an approach based on Technology Justice might look like and to highlight existing opportunities to make that actually happen.

Rebooting access

Agreeing the social foundation

A first step to rebooting our approach to achieving universal access to critical technologies has to be gaining broad acceptance of the concept and nature of a social foundation (and a global compact to deliver it), combined with a greater understanding and awareness among policymakers of how access to technology for basic services underpins the creation of that foundation.

The Sustainable Development Goals (SDGs) clearly offer an opportunity to pursue these issues, given they are a set of objectives that cover most of the bases of a social foundation (see Figure 13.2). At the time of writing, the indicators that will be used to measure progress against the goals themselves and their targets are still being negotiated. Their content is paramount in determining whether the SDGs will stand a chance of pushing global efforts on development and climate in the right direction. Although generalized references to technology abound or are strongly implied across many of the SDGs, at the more detailed level of the draft indicators, references to technology to date are somewhat sporadic. Draft indicators for SDG 2 (zero hunger) refer to improving agricultural extension services, the spread of more sustainable agricultural practices, and the maintenance of biodiversity among food crops and livestock, for example, whereas the draft indicators for SDG 6 (water and sanitation) and SDG 7 (affordable clean energy) make surprisingly little reference to technology in relation to access, particularly considering the debates around the role that off-grid infrastructure will need to play to ensure universal access (UN Stats, 2015).

According to Oxfam's Duncan Green, the SDGs are intended to influence:

- developing-country budgets and policies;
- wider social norms about rights and the duties of governments and others;
- aid volumes and priorities;
- developed-country budgets and policies (Green, 2015).

Figure 13.2 The Sustainable Development Goals
Source: UN, 2015

One big challenge with the SDGs, alongside the need to ensure they have useful and measurable targets and indicators, is determining how they can be used to exert real traction. In the case of developing-country budgets and policies, Green suggests this could be through:

- peer pressure, through countries internalizing the SDGs in domestic processes and acting as effective sources of pressure on their neighbours and others to follow suit;
- national media, as a source of pressure on decision-makers;
- civil society, in the form of campaigners and public educators;
- the private sector, as part of the change process (Green, 2015).

He suggests that the obvious questions that arise in response are:

- What kind of SDG reporting process or platform could help bring peer pressure about?
- What data and communications activity around the SDGs would be most likely to grab media interest at regular intervals over the next 15 years?
- What do civil-society organizations (particularly national ones) need in terms of access to data or support to make the SDGs an effective part of their advocacy work?
- What can be done to make the SDGs relevant to the interests of progressive national and international companies to either influence their operations directly or become part of the dialogue between business and government? (Green, 2015)

If the SDGs are to provide a truly transformative push towards the creation of a universally accessed social foundation, civil society in particular will have to play a major role in answering these questions and keeping the SDG agenda at the forefront of policy and media debates.

Sparking a data revolution

Rebooting the approach to access will also require a data revolution to properly understand the true scale of the task ahead and to track progress towards achievement of the social foundation for all. It is already known that there are critical flaws in much of the data collected in national and global statistics on access to basic services. These flaws can distort the view of existing levels of access, often inflating actual access figures. As work on the Sustainable Energy for All initiative's Global Tracking Framework has shown (see Chapter 3), simply counting connections to a service (for example, a water supply, electricity, or sanitation connection) is not enough; we need to know more about the quality and nature of the connection and whether people are actually able to meet their service needs as a result. Electricity 'access' figures, for instance, can reduce dramatically when the focus is shifted from simply counting connections to asking questions about the reliability and affordability of the supply and the impact of those factors on household lighting, cooking, heating, cooling, and communications.

The challenges are huge. In addition to the limitations of the targets of the SDG mechanism itself, there are limits to what a multisectoral global index can reasonably track before the big picture it is trying to communicate is drowned in detail. With 17 goals, 169 targets, and an as yet unknown but even larger set of indicators, this limit has probably already been breached. There are also real restrictions in the sourcing of meaningful data for such indexes. Official statistical data in many developing countries are weak or fail to capture all the information required. Supplementary data capture by mechanisms such as national household surveys following internationally standardized questionnaires and supported by the likes of USAID (Demographic and Health Surveys), The World Bank (Living Standard Measurement Surveys), UNICEF (Multi-Indicator Cluster Surveys), and some national censuses have been used to review energy access figures for the SE4All tracking framework (World Bank, 2015) and water and sanitation coverage for the UNICEF/WHO Joint Monitoring Project (JMP, 2015). But these supplementary questionnaires are themselves stretched and the capacity to add new questions to such surveys in order to collect additional data is extremely limited.

This is the problem the data revolution must address, not simply for global indicator sets such as the SDGs, but also for national ones. The revolution needs to go beyond (or at least supplement) traditional data collection methods, such as the census and household surveys, and embrace the opportunities offered by new technology and the way it is used. Greater use of satellite imagery, mobile phone data, crowdsourcing, smart metering, smart sensors, and data mining are all potential ways to source data, and are likely to engage not only traditional national statistics offices but also civil-society organizations and the private sector, as well as international organizations (Open Data Watch, 2015).

Increasing levels of transparency

A true data revolution would require this information to be made publicly available in easily accessible formats, for example, interactive maps that allow people to see how their village/district/region/country compares to others in terms of access to key services. This might be expressed in terms of the percentage of households with access to electricity, water, or sanitation services that meet defined standards in terms of quality, reliability, and affordability, or the percentage of households with access to the WHO list of essential medicines in their local public health clinic. Access to visually presented, often map-based, information has been shown, particularly in the water and sanitation sector, to be critical to helping marginalized communities engage with and lobby local and municipal authorities or national governments to change planning processes and rectify inequities in resource allocation to basic services (see, for example, the work of the Orangi Pilot Project in Pakistan (Welle, 2006) or the approach of Slum Dwellers International (SDI, 2015)). Accurate and accessible data also helps governments themselves to make the political case for pro-poor investments to their own populations as well as the investment case to donors and development banks.

The need for greater transparency does not stop with national governments. We need to be able to see if public policy is resulting in more money flowing towards the right priorities. That includes increased funds (public and private) for the provision of access to critical, affordable, and pro-poor basic services as well as increased funding for access to technical knowledge. The latter would range from improved and much expanded agricultural extension services for smallholder farmers to 'sustainability conversions' for sector professionals – for example, more energy-sector planners trained in renewables, and more agricultural extensionists with knowledge of agroecological approaches. Increased transparency is also needed among key funders, such as the multilateral development banks and bilateral donors, around the proportion of their investments portfolio that is flowing towards the more challenging but necessary investments to promote equity or environmental sustainability (for example, off-grid as opposed to on-grid investments in energy, or support to smallholder production in marginal areas in agriculture rather than large-scale commercial farming in high-production areas).

Including the excluded

More and better data, made publicly available, will make it easier to identify technology access problems and facilitate the media and civil society's role to hold authorities to account for provision of services to all. But universal access to a minimum social foundation will only be achieved once those who are marginalized (the poor, women, people with disabilities, older people, groups excluded on the basis of ethnicity or religion, and so on) have a say in the design and delivery of services that actually meet their needs. Inclusion can range from the general identification of needs through participatory mapping and other techniques (SDI, 2015) and by local participation in the design of individual service-delivery initiatives (Majale, 2009), to engagement in broader participatory budgeting and planning processes with government (ELLA / Practical Action, 2011).

But inclusion involves more than participation in the design of services alone. It also means ensuring the costs of those services are affordable and fair. Affordability of new or improved services is often referred to in terms of how much people are paying already for an inferior service. For example, solar home-lighting systems are sold in Kenya on a payment by instalment basis over a one-year period, so that the cost of each instalment is no more than the average family would have otherwise spent on kerosene for lighting for the same period. While there is a compelling logic to this approach, as people are getting a hugely improved service for the same price, it does not mean that access to energy services for the poor has suddenly become affordable. Rather, it means that a better source of lighting has become affordable. A full range of basic energy services for lighting, heating, cooking, cooling, communications, and some productive use probably remains out of reach of the majority of the poorest communities in the developing world, with one study suggesting this would cost between three and seven times a typical poor household's spending on energy (see Appendix 3). Clearly, in such circumstances, the challenge is not simply to make an existing level of service better by providing more, improved services at the same price, for example replacing lighting by kerosene with cleaner light sources, but to work out how to close the affordability gap for access to a full basic set of energy services.

Reaching beyond technology transfer

Technological catch-up in developing economies cannot be achieved solely through the transfer of technology developed in higher income countries. Learning from innovation systems approaches shows us that ecosystems have to be built and absorptive capacity developed in order to utilize imported technology. This is achieved through myriad interactions between different national and international actors, including deliberate shaping and directional efforts by government. The literature on inclusive innovation further suggests that a focus exclusively on the transfer of technologies from the developed world overlooks the opportunities for local innovation and the possibilities of applying, or further developing, useful indigenous knowledge and practice.

Despite this, many of the international efforts to support technology access in the developing world – for example, current and proposed arrangements to finance low-carbon development (Byrne et al., 2011) or support the SDG process (Kenny and Barder, 2015) – are conceived principally as an issue of technology transfer. This has to change if a sustained improvement in access to technologies necessary for achieving a social foundation is to take place. Alongside this, international funding needs to be applied to develop national capabilities to absorb and utilize imported technology, and to innovate locally. This is likely to involve: building networks of diverse stakeholders who can work together on technology projects and programmes; encouraging the development of shared visions of technology development and use among relevant stakeholders; promoting experimentation with technologies and practices; and supporting the sharing of learning from research and experience (Ockwell and Byrne, 2015).

Rebooting use

Gaining consensus on managing the risk

Alongside gaining broad acceptance of the concept and nature of a social foundation and the role technology plays in achieving that, it is clear that a reboot of humanity's relationship with technology is also needed. Policymakers and consumers in the developed and developing world alike need a greater understanding of the role technology use plays in exceeding Rockström's planetary boundaries along with a better appreciation of the technology paths that might lead to stability and environmental sustainability. This needs to go beyond the superficial treatment of technology in public and some policy debates on climate change, for example, which contain simplistic references to technology transfer for low-carbon development, or point to genetic modification as the only technical solution for improving agricultural productivity. A deeper, more informed, and more honest public policy debate is needed around the different technology pathways that could be followed, their potential impact on our economic models, and their likely risks and benefits for different stakeholder groups.

To improve our management of the unforeseen risks of new technology, a rearticulation of the precautionary principle is needed that removes the fog of multiple meanings and use in different international treaties and applications, and which facilitates a uniform and useful application of the principle in practice.

Building the pressure to engage

Building consensus around the nature of the risks faced and an understanding of the options available to act does not, of course, in itself guarantee action. We therefore need to continue to build on and strengthen existing levers for change as well as creating new ones.

Civil-society campaigns around environment and development are obviously important here, and building broader alliances that cross the environment/development and the developed/developing nations divides will continue to be important in establishing a powerful voice for change and a counternarrative to conventional views and politics. That voice must, in turn, take the opportunity offered by the SDGs to hold an unwavering mirror up to the world that highlights and condemns any misalignment by governments or other institutions between stated support for the goals and actual behaviour, while at the same time giving credit and publicity to positive action that demonstrates what is possible. Whatever their shortfalls, the SDGs represent a very unusual opportunity – a globally agreed vision of how the world could be a different and better place. As such, they remain the best starting point for a mechanism to hold each other to account. But new creative approaches from civil society are also needed to lever change: one such initiative is Carbon Tracker, which educates major investors by addressing them in their own language of risk and return, to cause a 'realignment of capital markets with climate reality' (Carbon Tracker, 2015). In a judo-like move, it aims to alter the direction of travel of an opponent by using its own strength and weight against it.

Moving to new economies

Rebooting our relationship with technology implies a radical rethink of that relationship rather than incremental change to it. Justice in the governance of the environmental and social impacts of technology use speaks to radically different models of economies, models that move away from conventional measures and drivers of change, such as the growth of material consumption, and look instead to improvements in human wellbeing and environmental sustainability as the key markers of progress. Herman Daly's steady state economic model (Daly, 1996), the work of French President Sarkozy's Commission on the Measurement of Economic Performance and Social Progress (Stiglitz et al., 2008), or the New Economic Foundation's ecological macro-economic modelling (Dafermos et al., 2015) are all pointers to a new sustainable economics. Two approaches in particular pose questions about the role of technology in supporting or even driving the necessary change.

New economy 1: a third industrial revolution? The American economist and social theorist Jeremy Rifkin is a somewhat controversial figure. A bestselling author of around 20 books on the impact of technological changes on the economy, society, and the environment, and, among other things, an adviser on energy security and climate change to the European Commission for the past decade (European Commission, 2014), his writing has attracted its fair share of criticism.[1] Nevertheless, in his book *The Third Industrial Revolution* (2011), Rifkin paints a picture of how to build on recent technology development to enable a transition to a very different economic model. It has at its core a compelling vision, whether or not you are an admirer of all his work. Rifkin's 'third industrial revolution' is a revolution that could come about through a real commitment to renewable power and through the exploitation of the emerging 'internet of things' and 'big data' capability.[2] In its world, every building is a green micro-power plant and the electricity grid turns into an energy internet that can connect millions of small energy producers to exchange surplus power. An internet of things (the 'things' in this case being the power-generating units) and big data analytics (to understand supply and demand dynamics from the flow of data from sensors embedded in those units) would be needed to manage such a system in an efficient and effective way. One benefit would be the potential to construct national 'smart grids' in rural areas in the developing world from the bottom up, by a process of 'swarming' smaller systems together over time.[3] But the really interesting idea that stems from Rifkin's vision is the way decentralized production of renewable power could combine with two other technologies, the internet and 3D printing, to have a transformative effect on manufacturing and a redistribution of economic power. Although a technology still in its infancy, 3D printing can potentially put the capability of producing low-volume or even one-off runs of complex objects at low cost into the hands of individuals. The internet, meanwhile, provides two additional necessary supporting conditions. Firstly, it provides free access to intellectual and technical expertise and property through a growing collaborative commons of open-source CAD files for 3D printable objects (see Chapter 11). Secondly, it offers free or very low-cost access to markets (as witnessed by the growth of the likes of eBay and Alibaba). Suddenly, the need to invest in a production line to produce goods economically, or in a large marketing and sales force to sell them, starts to recede.

While people are still a long way from being able to manufacture their own cars or iPhones (and it may never make sense to do so), it is already possible to manufacture some spare parts for them. The potential is for consumers to become producers (or 'prosumers' in Rifkin's term) and for that production to take place closer to the point of consumption (reducing the energy use, emissions, and costs of transportation), to be based on renewable power, and, perhaps, to even use locally grown biodegradable cellulose-based feedstock (see, for example, van Wijk and van Wijk, 2015).

Set alongside an ever-expanding move to new forms of collaboration and sharing – education through open online courses, crowdsourcing of funds for investment, car sharing for taxis (for example, Uber), and home sharing for vacations (Airbnb and so on) – the central attraction of Rifkin's vision is a different economy, a hybrid of capitalist markets and collaborative commons. But he also envisions an essentially greener and more sustainable form of production that is more evenly distributed, reducing the need for rural to urban migration for economic opportunity, and moving the ownership of the means of production away from the dominance of large corporations. Rifkin's third industrial revolution, in essence, is a revolution based on Technology Justice. It is supported by a form of technology that is environmentally sustainable and which doesn't impinge on future generations' ability to make choices and live the lives they value. To an extent, it democratizes control of technology and, through that, production.

New economy 2: the circular economy model. The concept of a circular economy, as propounded by the likes of the Ellen MacArthur Foundation, provides another and, in some ways, more mainstream vision of an alternative economy with Technology Justice at its centre. Forms of production and consumption traditionally have been largely linear: materials are collected, goods are produced, then used and, at some point, discarded. Little of the energy or material inputs are recycled into new forms of production. In Europe, for example, the recapture of energy or raw materials from waste accounts for only 5 per cent of the original raw material value and there are significant inefficiencies in resource use: the average car remains unused and parked for 92 per cent of the time, 31 per cent of food is wasted, and the typical office is occupied only 35–50 per cent of the time during working hours, to illustrate (Ellen MacArthur Foundation, 2015).

A circular economy, by contrast, seeks to decouple economic development from continued increase in consumption of what is a globally finite stock of natural resources by 'keeping products, components, and materials at their highest utility and value at all times' (Ellen MacArthur Foundation, 2015). This is achieved by following three principles (see Figure 13.3):

1. Preserve and enhance natural capital by choosing processes that use renewable resources or use resources more efficiently than others, and by delivering services virtually rather than physically wherever possible (for example, electronic music or books rather than hard copies). It also means looking for opportunities to regenerate resources, for example, by adding nutrients back into soil.
2. Optimize resource yields by designing products, and the components and materials that constitute them, so they can be circulated in the economy at their 'highest value of utility'. This means making products that are easy to repair, maintain, or, perhaps, upgrade, as this retains the most value already embedded in the product, including the energy already used to manufacture it. An example would be the Fairphone – a mobile phone designed to be easy to repair and upgrade to extend its life (Guvendil, 2014). This also means thinking about a product's component parts when it comes to recycling, the next level of utility down, where further energy and, perhaps, materials will be required to recombine constituents into new products. This would prioritize the use of pure materials wherever possible so that they can be put to other purposes after recycling without the need for further energy or chemical processes to separate them out. These processes apply to biological as well as technical cycles and require thought about how biochemical products can be reused or reintroduced into the biosphere in a way that helps to regenerate or restore them.
3. Foster system effectiveness by trying to understand the externalities of the production process which are generally ignored in financial or economic analysis – potential negative impacts on the environment or health, for example – and then trying to redesign those processes to remove or minimize the negative impacts.

Given that the primary purpose of the circular economy approach is, through product and process design, to minimize or eliminate negative environmental or social impacts, it embeds at its core the principle of Technology Justice – the tenet that everyone should have the right to use technologies that help them live the life they value, provided that doesn't impact on others' ability now or in the future to do the same.

In addition to environmental benefits, circular economies have the potential to create substantial employment benefits both in the developed and the developing world, with the potential to impact on poverty in the latter. Indeed, in the developing world informal workers already play an

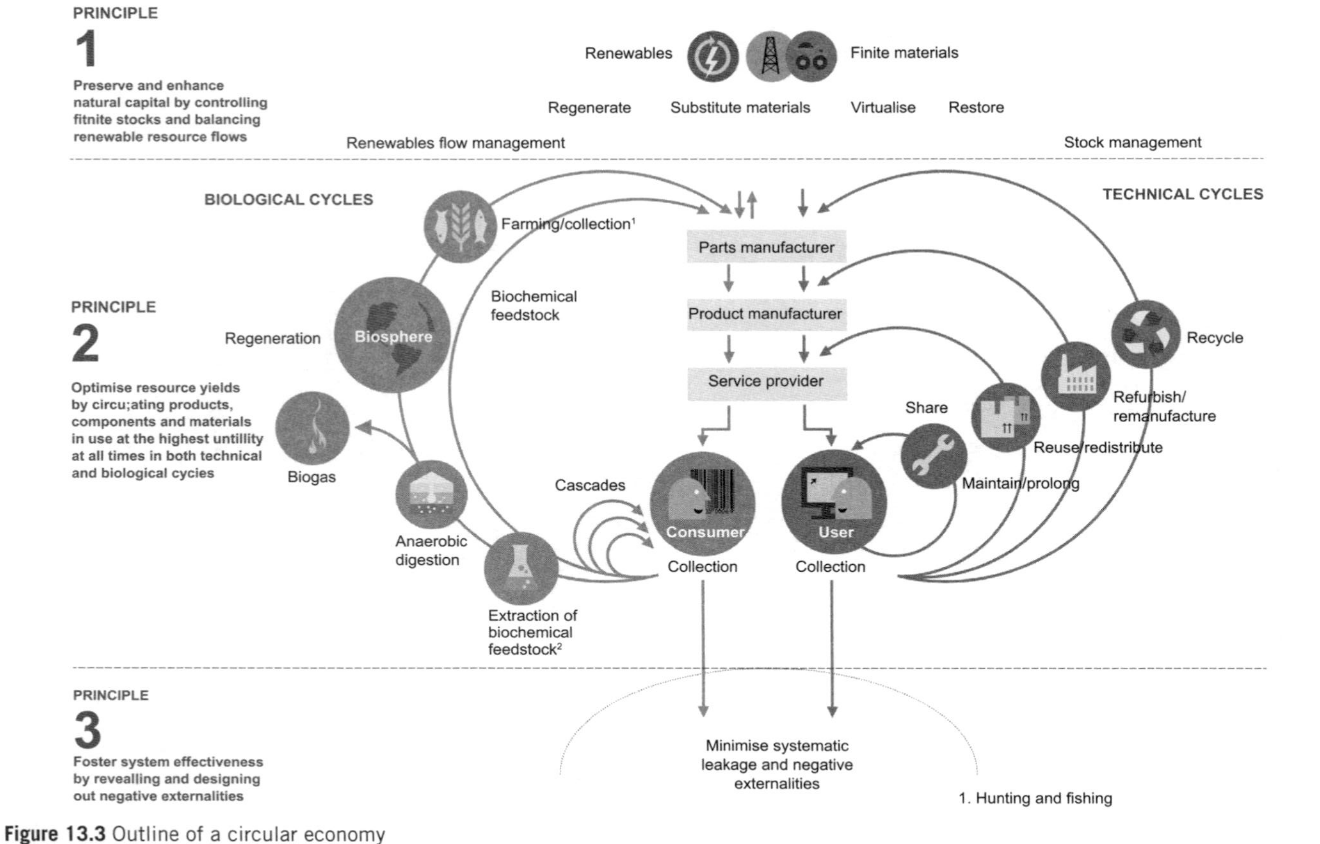

Figure 13.3 Outline of a circular economy

Source: Ellen MacArthur Foundation, 2015: 6

important part in creating, sustaining, and growing circular economies, through networks of rebuilding, refurbishing, fixing, and augmenting technologies in frugal and innovative ways. It has been estimated that around 1 per cent of the developing world's urban population survives by salvaging recyclables from waste (Medina, 2008). Although much of this involves the poorest of the poor separating plastics, metals, and other recyclable matter from rubbish to sell on, it can also involve artisanal work ranging from the making of shoes from old car tyres to tin smiths beating oil drums into cooking stoves, to mechanics salvaging and repairing parts to keep vehicles on the road. Recycling also supports jobs in the formal sector in the developing world. The company Protoplast in Senegal produces some 15 tons of plastic resins a month from waste plastic collected locally (Lemoigne, 2015), while the social enterprise Protoprint in Pune, India, produces filament for 3D printers from a similar source (Marks, 2014). Ethical procurement policies such as Protoprint's and access to improved tools and equipment can help improve working conditions, efficiency, and incomes of poor waste pickers as well as creating jobs in the 'upcycling' manufacturing process itself. A wide range of policy interventions, such as green procurement policies on behalf of government or legislation that limits the amount of non-recyclable packing in products, have the potential to expand the market and its related employment opportunities further (Liebenberg, 2007).

Rebooting innovation

Responsible innovation

The final component of a reboot of the relationship with technology is the oversight of the direction of the collective scientific and technological endeavour. Governance systems need to be able to do more than ensure that the dissemination and use of existing technology manoeuvres humanity into that safe and just space between a universally achieved social foundation and environmental planetary boundaries. They need also to ensure that the global technological innovation effort improves our ability to stay in that safe space as population increases and new, as yet unanticipated, challenges emerge. Technology Justice requires a governance system that goes beyond trying to understand and mitigate the potential risks associated with the development of new technology to questioning the very purpose of innovation and ensuring the collective effort is channelled in the most effective and efficient way to address current issues of poverty and environmental sustainability. This will require the adoption of some of the governance techniques being developed by supporters of responsible innovation, including the four principles of anticipation, reflexivity, inclusion, and responsiveness (see Chapter 11).

Understanding and building innovation systems in the developing world

As the discussion in Chapter 7 showed, innovation, whether the development of a new technology or the adoption of an existing one in a place where it has not been used before, is dependent on a broad network of actors and institutions. The simple act of transplanting technologies from the developed to the developing world rarely works without an understanding of the physical, institutional, and social environment into which it is being transplanted. The concept of national innovation systems has been well researched and applied in the developed world to account for this, but remains under-researched and poorly understood in developing-country contexts, where innovation systems may be fragmented and still evolving. Account needs to be taken in technology-transfer processes of the fact that developing-country innovation systems are often substantially different from the mature innovation systems found in developed economies and that simply imitating innovation policies practised in developed countries is unlikely to deliver the expected results. Given the cost of technology transfer to the developing world to meet poverty and climate goals is likely to reach hundreds of billions of dollars a year,[4] the absence of research into the nature of existing innovation systems in developing countries, and how they can be strengthened to support the envisaged level of technology transfer, is a massive knowledge gap that must be urgently addressed.

Improving global coordination

The reboot required is not only radical but also urgent. A very small window of opportunity is left to address and mitigate climate change before natural feedback loops potentially come into play which could set a self-sustaining process of warming in place that would render further human intervention to reduce emissions meaningless. This urgency requires resources to be marshalled and applied in the most effective way possible. At the very least this needs global agreement on the most essential areas for further science and technology research and innovation, to help the world stay within Rockström's planetary boundaries and to deliver Raworth's social foundation. The establishment of the Global Health R&D Observatory in 2012 by the World Health Council is an attempt to do just this for the health sector (see Chapter 8) and shows that such an endeavour is not impossible. Moving beyond identifying needs to pooling resources and coordinating research efforts has proven a more elusive goal, but is exactly what global mechanisms such as the observatory or the new SDG Technology Facility and the UNFCCC Technology Mechanism will have to focus on. That will require real transformation of not only those bodies but also of the thinking and actions of the member states that constitute the institutions which oversee them.

Supporting open and inclusive innovation systems

Existing proprietorial systems of innovation focused on the protection of intellectual property rights and patents are not driving technological innovation in the areas that most matter because they are the areas with the weakest market signals. What is more, although secrecy and competition may produce results where resources and time are both abundant, those luxuries no longer exist. Wasting resources on companies and research institutions potentially pursuing research pathways that others have already determined are dead ends is not an option. Instead, there is a need to learn and borrow from the experience of open-source innovation to build quickly on the cumulative experience of past research rather than having to wait for one individual institution to deliver a particular breakthrough on its own. Innovation processes also need to be inclusive. Engagement, for example, of researchers in developing countries in global efforts has the added value of building national technical capabilities and opening up research efforts to the potential of learning from many sets of dispersed indigenous knowledge.

Rethinking the relationship between public and private sectors

Finally, the existing public policy narrative on the roles the state and the private sector play in the innovation process has, itself, to be rebooted. Both sectors are critical. But the current narrative fails to recognize the real entrepreneurial role the state plays in innovation, not just in the critical area of public-funded basic research, but also with regard to public investment in supporting nascent industries before venture capital or other private finance steps in.

The challenge is to go beyond seeing the rhetoric change from a portrayal of the state as a lousy picker of winners to an active entrepreneurial risk-taker. Higher expectations have to be imposed on the private sector. The financialization of the reward system in the private sector over the past decades has to be countered, along with the parasitic behaviour it engenders, by seeking taxation systems aligned instead to the promotion of real innovation resulting in beneficial social and environmental impact, and the penalizing of the simple value-extraction grab-and-run mentality that has characterized both corporate and venture capital fund behaviour. Given the state's entrepreneurial activity and the role it should play in enacting the principles of responsible innovation and guiding the purpose of innovative efforts, it is only right that consideration be given as to how to improve the state's return on investment in order to ensure it can adequately finance these responsibilities.

Changing the way technology is governed

The broad governance challenge

Across all three strands of technology access, use, and innovation discussed in this book there has been a call for a change in the way governance systems work. The term 'governance' has been used in a broad fashion to refer not just to the actions of governments and international intergovernmental institutions, but also to the efforts of all the individual and institutional actors involved in the development, take-up, and use of technology, and to all the coordination and mutual accommodation between those agents that this involves. The system under consideration is thus vast and hugely complex.

As Walter Valdivia, a fellow of the US-based Brookings Centre for Technology Innovation, says:

> The idea of governance of socio-technical systems is daunting. How do we even begin to understand it? What kinds of modes of governance exist? What are the key dimensions to understand the integration of socio-technical systems? And perhaps more pressing, who prevails in disputes about coordination and accommodation? (Valdivia, 2016)

Valdivia points to the work of Susana Borrás and Jakob Edler (2014) as a guide to how the governance of social-technical change can be understood. Three key analytical dimensions emerge from their case studies:

1. Who drives change? Whether the emphasis is on individual actors self-organizing or on mission-orientated institutions, such as governments, providing the lead.
2. How is change engineered? Whether the emphasis is on law and regulatory instruments, on collaboration through networks to a common goal (think of open-source approaches), or through markets condensing all the necessary information for rational decision-making into price signals.
3. What gives change legitimacy and leads to acceptance by society? How, for example, is the tension between the role of citizens and that of scientific experts in the decisions about collective problems and their solutions resolved? (Borrás & Adler, 2014)

This book has consistently argued that, given the urgency of the challenges faced, the project of finding a safe, inclusive, and sustainable space for humanity requires strong mission-orientated leadership. Change would be engineered through a combination of regulation and collaborative work to a common goal, with a lessening of the emphasis on a leading role for market-based decision-making, given the latter's track record of driving behaviour in the wrong direction. Legitimacy has to come from greater engagement of citizens in debates around the direction of science and technology innovation and how the risks associated with technology innovation and deployment are managed. This has to be combined, particularly in the developing world, with engagement with the marginalized populations who lack access to basic levels of technology in decision-making around planning and budget priorities.

The mission-orientated leadership has to happen at both international and national levels. In systems, as Valdivia points out, when an individual actor adapts and changes, their actions trigger further adaptations by others in the system to accommodate those changes. Given the complexity and scale of technology governance systems, the challenge will be to concentrate on the behavioural change of key actors who, by adapting, will have the greatest knock-on effect on others in those innovation systems.

The particular governance challenges of global institutions

At a global level there are several institutional challenges that require urgent attention if a working governance system for technology is to be found. Without these changes, progress at a national level will be very difficult.

The World Trade Organization's TRIPS mechanism is among the most significant international regulatory instruments in existence, but its role has been corrupted. It has become the vehicle for powerful transnational corporations to threaten legal action on grounds of restraint of trade against states attempting to enact environmental regulation, or the means by which the financial gains or returns on technology development are enclosed and privatized through the patent system, while the associated risks are left to be borne by the public. In reality, an international convention and institution is needed that is able to exert the same level of pressure but in the opposite direction – to restrict international trade in goods and services that are inherently socially or environmentally unsustainable and to force an opening up of intellectual property that is deemed critical to meeting environmental or developmental goals.

Alongside mitigating the negative role played by the WTO, other international institutional changes are required for Technology Justice. Multiple UN-led international processes have taken on technology coordination roles as part of their mandates. As well as the SDG Technology Facility Mechanism and Technology Bank, the UNFCCC's Subsidiary Body for Scientific and Technological Advice (SBSTA), Technology Executive Committee (TEC), and Climate Technology Centre and Network (CTCN) have these roles. Even the UN's office for disaster risk reduction (UNISDR) has a Scientific and Technical Partnership under the Sendai Framework which aims to '(e)nhance the scientific and technical work on disaster risk reduction and its mobilization through the coordination of existing networks and scientific research institutions at all levels and all regions' (UNISDR, 2015). Apart from the obvious need for better coordination to deal with the confusion of overlapping mandates, and the resultant inefficient use of resources by this plethora of institutions, there are big gaps in their terms of reference that need to be addressed urgently (see, for example, UN-OHRLLS, 2013, 2015; UN Sustainable Development Knowledge Platform, 2015; UNFCCC, 2015a, b), in particular:

- In terms of technology innovation, there is a complete absence of any mandate to identify science and technology research and development priorities or to coordinate global approaches to R&D in areas where technological advance is vital.
- In terms of technology use, there is no mechanism for identifying, assessing, and acting on the risks posed by the use of existing or new technologies to poverty or environmental goals. For example, there is no mechanism under the UNFCCC's SBSTA, TEC or CTCN that would allow the risks of technologies proposed for geoengineering approaches

to climate control to be weighed or their use regulated. Likewise, under the SDG Technology Facilitation Mechanism, there is no mandate to look at the relative risks and advantages of technologies for alternative approaches to, for example, food production within Goal 2 (zero hunger), such as conventional high-input agriculture, 'sustainable intensification' techniques, or agroecological approaches.

- In terms of technology access, across both the UNFCCC and SDG mechanisms, there is no commitment to fill the gap in research on how national innovation systems work in developing countries and the factors governing successful technology transfer, a knowledge gap that puts trillions of dollars of projected investment in technology transfer at risk. There is also no commitment to specifically monitor the success of the technology transfers that are promoted under the mechanisms at the moment. Finally, in the case of the SDG Technology Bank, a business-as-usual approach to intellectual property rights is being promoted, an approach that seeks to integrate developing countries into the existing global patents system instead of driving the radical change needed to stimulate innovation in vital technologies for poverty relief and environmental sustainability, where market signals remain weak.

The mandates of the technology facilities and mechanisms established under the SDG and UNFCCC processes are confused, their power is weak, and their resource base very small. They need strengthening politically, they need a clearer, wider, and better coordinated mandate, and they need a massive increase in financial and human resources to do the job required. The latter point is highlighted by the contrast between the resources available to these mechanisms compared to the WTO. The WTO has over 630 staff on its regular budget (WTO, 2014: 136), compared to just '4–7' part-time staff anticipated for the secretariat of the UN Interagency Task Team on Science, Technology and Innovation for the Sustainable Development Goals, the body that will support the Technology Facilitation Mechanism (UN Sustainable Development Knowledge Platform, 2015: 2). Likewise, the WTO has an annual budget of approximately $200 m (WTO, 2014: 144), compared to around $112 m[5] for the UNFCCC in its entirety, of which just $700,000[6] is allocated for its technology oversight activity through the SBSTA, TEC, and CTCN.

Today, we have put an order of magnitude more resources into the organization charged with restricting and licensing the flow of technical know-how around the world than is allocated to supporting the institutions charged with ensuring the technological advances necessary to meet critical poverty and environmental goals. This has to change.

Technology as if people and planet mattered

The purpose of this book is to explore, through the lens of technology, the two great challenges for humanity today: ending poverty and securing a productive and stable environment to support future generations. The intent is to highlight a dilemma – how access to technology is critical to human development but how the use of technology threatens human survival – and to suggest that rebooting our relationship with technology by using Technology Justice as a guiding principle of governance might help reconcile what at first may seem irreconcilable.

Many of the tools needed to do this are already to hand. The majority of the technologies needed to establish a minimum social foundation already exist. So do some of the approaches that could help drive the collective innovative effort in the right direction, such as responsible research, open-source innovation, or circular economics. Even the global agreements and institutions that will be vital to setting overarching goals, coordinating efforts, and arbitrating agreements on technology are at least partly in place. A more radical version of the Technology Facilitation Mechanism and the Technology Bank could well provide the necessary impetus and oversight to achieve what needs to be done, as a case in point.

These tools and institutions exist, but they are not being used, which is why we need a profound and swift change to the way we govern the development, dissemination, and use of technology. That new governance approach has to be one that considers technology dissemination, use, and innovation as if people and planet actually mattered.

We have the tools to hand. Technology Justice is possible. But only if the necessary and radical practical action is taken now.

Notes

1. The evolutionary biologist Stephen Jay Gould reviewed one of Rifkin's early books on genetics in an article entitled 'On the origin of specious critics', describing the book as 'a cleverly constructed tract of anti-intellectual propaganda masquerading as scholarship' (Gould, 1985). The rigour of the thinking underpinning his latest book, the *Zero Marginal Cost Society* (2014), has also received mixed reviews (see, for example, Walters, 2014; Ogden, 2014).
2. The 'internet of things' refers to the idea that large numbers of sensors embedded in physical objects, machines, and the environment could share data across the internet, while 'big data' refers in this case to the analytical power to be able to turn such data feeds into useful information that can be acted on (see, for example, Idle, 2015).
3. The process of 'swarming' has been proposed as a means of building national grids out from urban centres to unserved rural areas in the developing world. Swarming would involve enhanced take-up of solar home systems leading to individual homes and institutions being linked together to form local mini-grids which, over time, would be further linked into ever larger grids until the resulting network joins and forms part of a smart national grid (see, for example, Philipp et al., 2014).

4. As mentioned in Chapter 7, estimates suggest the requirements for mitigation finance alone, much of which would be needed for the transfer of clean energy technologies, could run to $265–565 bn per annum for the next 20 years (Montes, 2013).
5. Based on the summation of figures from the proposed programme budget for the biennium 2016–17 (UNFCCC, 2015c: 14) and the 2016–17 addendum, 'Activities to be funded from supplementary sources' (UNFCCC, 2015d: 4).
6. Based on table items 17 and 20 from the 2016–17 addendum (UNFCCC, 2015 d: 4).

References

Borrás, S. and Adler, J. (2014) *Governance of Social-Technical Change: Understanding Change*, Cheltenham: Edward Elgar.

Byrne, R., Smith, A., Watson, J., and Ockwell, D. (2011) *Energy Pathways in Low Carbon Development: From Technology Transfer to Socio-Technical Transformation*, Brighton: STEPS Centre, Institute of Development Studies.

Carbon Tracker (2015) 'What is Carbon Tracker?' <http://www.carbontracker.org/> [accessed 3 December 2015].

Daly, H. (1996) *Beyond Growth*, Boston, MA: Beacon Press Books.

ELLA / Practical Action (2011) *Participatory Budgeting: Citizen Participation for Better Public Policies*, Rugby: Practical Action Publishing.

Ellen MacArthur Foundation (2015) 'Towards a circular economy: business rationale for an accelerated transition', London: Ellen MacArthur Foundation.

European Commission (2014) 'Jeremy Rifkin – plenary – digital action day 2014' [online], <https://ec.europa.eu/digital-agenda/en/jeremy-rifkin-plenary-digital-action-day-2014> [accessed 4 December 2015].

Green, D. (2015) 'From Poverty to Power: Hello SDGs, what's your theory of change?' *Oxfam blogs* (posted 29 September) <http://oxfamblogs.org/fp2p/hello-sdgs-whats-your-theory-of-change/> [accessed 2 December 2015]

Gould, S. (1985) 'On the origin of specious critics', *Discover*, January: 34–42.

Guvendil, M. (2014) 'Next step in life cycle assessment', 20 June, *Fairphone*, <https://www.fairphone.com/2014/06/20/next-step-in-life-cycle-assessment-inventory-analysis/> [accessed 4 December 2015].

Idle, T. (2015) 'Using the power of big data to disrupt and transform today's energy market', 15 December, *Virgin*, <https://www.virgin.com/virgin-unite/leadership-and-advocacy/using-the-power-of-big-data-to-disrupt-and-transform-todays-energy> [accessed 26 January 2016].

JMP (2015) 'Definitions, methods and data sources', *UNICEF/WHO Joint Monitoring Programme*, <http://www.wssinfo.org/definitions-methods/data-sources/> [accessed 2 December 2015].

Kenny, C. and Barder, O. (2015) 'Technology, development, and the post-2015 settlement', CGD Policy Paper 63, Washington, DC: Centre for Global Development.

Lemoigne, R. (2015) 'Can emerging countries benefit from the circular economy?', 23 September, *Circulate*, <http://circulatenews.org/2015/09/

can-emerging-countries-benefit-from-the-circular-economy/#_ftnref6> [accessed 28 January 2016].

Liebenberg, C. (2007) 'Waste recycling in developing countries in Africa: barriers to improving reclamation rates', presented at *Eleventh International Waste Management and Landfill Symposium*, Sardinia.

Majale, M. (2009) *Developing Participatory Planning Processes in Kitale, Kenya*, Nairobi: UN Habitat.

Marks, P. (2014) 'Ethical 3D printing begins with plastic waste pickers', 16 July, *New Scientist*, <https://www.newscientist.com/article/mg22329784-200-ethical-3d-printing-begins-with-plastic-waste-pickers/> [accessed 28 January 2016].

Medina, M. (2008) 'The informal recycling sector in developing countries', *Gridlines Note 44*, Washington, DC: The World Bank.

Montes, M.F. (2013) 'Climate change financing requirements of developing countries', *Climate Policy Brief 11*, Geneva: South Centre.

Dafermos, Y., Galanis, G., and Nikolaidi, M. (2015) 'A new ecological macroeconomic model: Analysing the interactions between the ecosystem, the financial system and the macroeconomy', London: New Economics Foundation.

Ockwell, D. and Byrne, R. (2015) 'Improving technology transfer through national systems of innovation: climate relevant innovation-system builders (CRIBs)', *Climate Policy* <http://dx.doi.org/10.1080/14693062.2015.1052958>.

Ogden, T. (2014) 'No value', *The Stanford Social Innovation Review*, <http://ssir.org/book_reviews/entry/no_value> [accessed 4 December 2015].

Open Data Watch (2015) 'Data for development: an action plan to finance the data revolution for sustainable development', New York: Open Data Watch / Sustainable Development Solutions Network.

Philipp, D., Kirchhoff, H., Edlefsen, B., and Theune, J. (2014) 'Swarm electrification – a paradigm change: building a micro-grid from the bottom-up', 15 August, *Energypedia*, <https://energypedia.info/wiki/Swarm_Electrification_-_A_Paradigm_Change:_Building_a_Micro-Grid_from_the_Bottom-up> [accessed 28 January 2016].

Rifkin, J. (2011) *The Third Industrial Revolution*, New York: Palgrave Macmillan.

Rifkin, J. (2014) *The Zero Marginal Cost Society*, New York: Palgrave Macmillan.

SDI (2015) 'Know your city', *Slum Dwellers International*, <http://www.knowyourcity.info/map.php#/app/ui/world> [accessed 2 December 2015].

Stiglitz, J., Sen, A., and Fitoussi, J. (2008) *Report by the Commission on the Measurement of Economic Performance and Social Progress*, Paris: Commission on the Measurement of Economic Performance and Social Progress.

UN (2015) 'Sustainable development goals: 17 goals to transform our world', *United Nations*, <http://www.un.org/sustainabledevelopment/sustainable-development-goals/> [accessed 2 December 2015].

UNFCCC (2015a) 'Technology executive committee mandates, functions, modalities and rules of procedure', *UNFCCC TT Clear*, <http://unfccc.int/ttclear/templates/render_cms_page?s=TEC_mandates> [accessed 29 January 2016].

UNFCCC (2015b) 'About the CTCN', *UNFCCC TT Clear*, <http://unfccc.int/ttclear/templates/render_cms_page?TEM_ctcn> [accessed 29 January 2016].

UNFCCC (2015c) 'Proposed programme budget for the biennium 2016–2017, note by the Executive Secretary', Geneva: UNFCCC.

UNFCCC (2015d) 'Proposed programme budget for the biennium 2016–2017, note by the Executive Secretary, Addendum: Activities to be funded from supplementary sources', Bonn: UNFCCC.

UNISDR (2015) 'Terms of reference of the scientific and technical partnership for the implementation of the Sendai framework for disaster risk reduction 2015–2030', Geneva: United Nations.

UN-OHRLLS (2013) 'A technology bank and science, technology and innovation supporting mechanism for the least developed countries', informal background note, New York: UN.

UN-OHRLLS (2015) *Feasibility Study for a United Nations Technology Bank for the Least Developed Countries*, New York: UN.

UN Stats (2015) *Results of the List of Indicators Reviewed at the Second IAEG-SDG Meeting, 2nd November 2015*, New York: UN.

UN Sustainable Development Knowledge Platform (2015) 'Terms of reference for the UN Interagency Task Team on Science, Technology and Innovation for the Sustainable Development Goals', <https://sustainabledevelopment.un.org/content/documents/8569TOR%20IATT%2026%20Oct%202015rev.pdf> [accessed 25 January 2016].

Valdivia, W. (2016) 'Why should I buy a new phone? Notes on the governance of innovation', 28 January, *Brookings Tech Tank – improving technology policy*, <http://www.brookings.edu/blogs/techtank/posts/2016/01/22-governance-of-innovation-book-review-valdivia?rssid=TechTank> [accessed 30 January 2016].

van Wijk, A., and van Wijk, I. (2015) *3D Printing with Biomaterials*, Amsterdam: IOS Press / Delft University Press.

Walters, R. (2014) '*The Zero Marginal Cost Society* by Jeremy Rifkin', the *Financial Times*, <http://www.ft.com/cms/s/2/7713c7fc-b07a-11e3-8efc-00144feab7de.html> [accessed 4 December 2015].

Welle, K. (2006) *Water and Sanitation Mapping in Pakistan*, London: WaterAid.

World Bank (2015) 'Progress toward sustainable energy – global tracking framework 2015', *The World Bank*, <http://trackingenergy4all.worldbank.org/data> [accessed 2 December 2015].

WTO (2014) *Annual Report 2014*, Geneva: World Trade Organization.

EPILOGUE

Is small beautiful?

In the Prologue I referenced E.F. Schumacher's 1972 book *Small is Beautiful* as a source of inspiration. Schumacher, famously, did not like his book's title, which was dreamed up by the publisher. Although Schumacher argued in the book that we need to rethink our ideas of economies of scale, he wasn't actually saying that *everything* had to be small, rather that we shouldn't conversely try to push everything to scale. In parts he seems almost exasperated about this:

> What I wish to emphasise is the duality of the human requirement when it comes to the question of size: there is no single answer. For his different purposes man needs many different structures, both small ones and large ones, some exclusive and some comprehensive ... (Schumacher, 1973: 48)

We do need to organize ourselves in some ways at a global level: dealing with conflict, poverty, environmental degradation, and sustainable and fair use of natural resources all require us to act on a scale that crosses national boundaries. But Schumacher's warning still stands today. We also need to rethink the idea of scale in relation to economics and technology, as the trend to organize our economies around ever grander structures has increased rather than decreased the gap between the rich and the poor, and led to pursuit of an economic model which is clearly unsustainable.

The challenge is to grasp the complexity of the task ahead and to avoid reaching for a simple single solution for everything where none actually exists. We need to think both small and big, depending on the job at hand. Schumacher's response to this dilemma was to focus on 'smallness' as the underdog:

> Today, we suffer from an almost universal idolatry of giantism. It is therefore necessary to insist on the virtues of smallness – where this applies. (If there were a prevailing idolatry of smallness, irrespective of subject or purpose, one would have to try and exercise influence in the opposite direction). (Schumacher, 1973: 48)

Smallness remains the underdog today but, importantly, new tools and technologies are emerging that have the potential to change this.

At the heart of *Small is Beautiful* is the idea of returning to a form of technology that is more human in scale and less damaging to the environment.

> The technology of mass production is inherently violent, ecologically damaging, self-defeating in terms of non-renewable resources and stultifying for the human person. The technology of production by the masses, making use of the best of modern knowledge and experience,

http://dx.doi.org/10.3362/9781780449043.015

> is conducive to decentralisation, compatible with the laws of ecology, gentle in its use of scarce resources, and designed to serve the human person instead of making him the servant of machines. I have named it intermediate technology to signify that it is vastly superior to the primitive technology of bygone ages but at the same time much simpler, cheaper, and freer than the super-technology, or democratic or people's technology – technology to which everybody can gain admittance and which is not reserved to those already rich and powerful ... (Schumacher, 1973: 126)

Intermediate technology does not feature much in today's literature and policy debates around technology innovation and technology transfer needs. But technology development in the years since *Small is Beautiful* was published has only made the need to rethink technology and scale more urgent. The agricultural sector is a clear example of this, where the industrialization of food production has led to a massive increase in the amount of energy used per unit of food produced, widespread pollution of the environment from fertilizer and pesticide use, and a loss of genetic variety in livestock and crops which puts the future of our global food supply in jeopardy. Another example is the impact of the rapid scaling up of automation on whole classes of jobs and employment.

Interestingly, though, while developments over the past 40 years have made the reintroduction of human scale into technology even more urgent, they have made it more possible, too. Innovation in solar photovoltaics has made household-level power production, both for self-consumption and sale, not just possible but also economic. We can all now become independent power producers. Improvements in information and communications technologies, including the internet and mobile telephony, have opened up free access for individuals to vast amounts of technical knowledge and also provided a marketing mechanism for small-scale producers that used only to be in the hands of large corporations with massive advertising budgets. The introduction of 3D printing will further revolutionize what can be produced locally and on a small scale. Meanwhile, better understanding of agroecology offers new opportunities to improve the competitiveness and economics of small-scale food production while addressing environmental risks.

The need for us to organize on a scale that crosses national boundaries to deal with conflict, poverty, environmental degradation, and sustainable and fair use of natural resources has not diminished. And there will still be forms of production and economic activity that lend themselves to large structures. But technology development, combined with an increasing interest in collaborative commons and open-source approaches to innovation, puts new tools into the hands of individuals and opens up possibilities for small-scale, local production, the sum of which has the potential to make a difference at scale. Some of those options will also be significant in addressing environmental issues, rural to urban population movements, and patterns of democratic power and control of resources.

Human wellbeing is about more than just access to material goods and services. Wellbeing also requires a sense of being in control of one's own life and destiny and of having a say in decisions that impact on that ability. Today we increasingly have technological options that allow us to choose to return ownership of some of the means of production back to individuals and, in so doing, strengthen this aspect of wellbeing. Where such options are also likely to deliver real and significant poverty and environmental benefits, we should take them. In that sense, small is still beautiful.

Reference

Schumacher, E.F. (1973) *Small is Beautiful: A Study of Economics as if People Mattered*, Oxford: Blond and Briggs.

APPENDIX 1

Failures to adhere to the precautionary principle

Technology	*Warning signs first raised*
Lead in petrol	The neurotoxic effects of lead were recognized as far back as Roman times. In 1925, at the 'one-day trial' of leaded petrol in the US, many experts warned of the likely health impacts of adding lead to petrol. Yet, despite the availability of an equally effective alcohol additive which was assessed by experts to be cleaner, the leaded route to fuel efficiency was chosen in the US and then exported to the rest of the world. Phased out from the mid-1980s. Most countries stopped using by 2012.
PCE (perchlor-ethylene), used in the production of plastic linings for drinking-water distribution pipes in the late 1960s and 1970s	PCE had been used to treat hookworm and the literature contained data on side effects. Later, a variety of occupational users were studied, including aircraft workers, small companies in countries where biological monitoring was required, and dry-cleaning companies. Several environmental studies were also conducted to see if drinking water contaminated with PCE or its close relative, TCE (trichloroethylene), was associated with cancer. Results were mixed and the chemical industry consistently denied that PCE was a human carcinogen. In the early 1970s it was confirmed that PCE was cacogenic and in 1976 it was discovered that PCE had been leaching into the water from the pipe linings, causing widespread contamination of water supplies that today still require continuous remediation. Environmental and occupational standards were promulgated, but continued to generate controversy as the chemical industry attempted to create doubt in the minds of decision-makers over the interpretation of scientific evidence. PCE has since figured in lawsuits.
Methylmercury used in the production of acetaldehyde	Methylmercury can cause Minamata disease, which can induce lethal or severely debilitating mental and physical effects. The disease came to prominence in the 1950s in populations around Minamata Bay in Japan. It was officially identified in 1956 and attributed to factory effluent but the government took no action to stop contamination or prohibit fish consumption. Chisso, Japan's largest chemical producer, knew it was discharging methylmercury and could have known that it was the likely active factor but chose not to collaborate and actively hindered research. The government concurred, prioritizing industrial growth over public health. In 1968, Chisso stopped using the process that caused methylmercury pollution and the Japanese government then conceded that methylmercury was the etiologic agent of Minamata disease. Between 1932 and 1968, 488 tonnes of mercury had been dumped in the sea. It was not until 2009 that UNEP initiated a global mercury phase-out, with a global legally binding instrument on mercury signed in 2013.

Technology	*Warning signs first raised*
Beryllium, used in manufacturing processes, including the production of nuclear weapons	Over several decades, increasingly compelling evidence accumulated that chronic beryllium disease (CBD, an irreversible inflammatory lung disease) was associated with beryllium exposure at levels below the existing regulatory standard (2.0 µg/m^3, adopted in the US for weapons manufacturers in 1949 and more widely for industry in 1971). The beryllium industry had a strong financial incentive to challenge the data and decided to be proactive in shaping interpretation of scientific literature on beryllium's health effects. It hired public relations and 'product defence' consulting firms to refute evidence that the standard was inadequate. When the scientific evidence became so great that it was no longer credible to deny that workers developed CBD at permitted exposure levels, the beryllium industry responded with a new rationale to delay promulgation of a new, more protective exposure limit. Standards of 0.2 µg/m^3 were set specifically for nuclear weapons and clean-up workers by the US Department of Energy in 1999. But in 2009 the American Conference of Governmental Industrial Hygienists recommended a far stricter exposure limit of 0.05 µg/m^3. As of 2012, however, the Occupational Safety and Health Administration in the US Department of Labor had yet to propose a new general workplace beryllium standard.
DBCP (dibromochloropropane), a pesticide	DBCP was introduced into US agriculture in 1955 and approved for use as a fumigant in 1964. It was used as a pesticide against nematodes (roundworms or threadworms) that damage pineapples, bananas, and other tropical fruit. By 1961, laboratory experiments had shown that it made the testicles of rodents shrink and significantly reduced the quantity and quality of sperm. Nonetheless, the compound was widely marketed and became a commercial success. In 1977, workers at a production plant became worried that they were unable to father children. An emergency study by a US government agency discovered that in many cases the workers were suffering from deficient or absent sperm. While controls were improved at US facilities, the product continued to be marketed and sprayed in Latin America, the Philippines, some African countries, and elsewhere. By the 1990s, tens of thousands of plantation workers in these countries had allegedly suffered adverse reproductive effects from DBCP use. The story continues today with contentious legal claims for compensation, contamination of drinking water, and industry attempts to prevent a Swedish documentary on the issue from being screened.
Seed-dressing systemic insecticides	In 1994, French beekeepers began to report alarming signs. During summer, many honeybees did not return to the hives. Honeybees gathered together in small groups on the ground or hovered, disoriented, in front of the hive and displayed abnormal foraging behaviour. These signs were accompanied by winter losses. Evidence pointed to Bayer's seed-dressing systemic insecticide Gaucho®, which contains the active substance imidacloprid. In January 1999, the Ministry of Agriculture decided to ban Gaucho® in sunflower seed-dressing for two years, applying the precautionary principle. Bayer challenged the ministerial decision in the administrative court of Paris, which eventually found in favour of the government. In 2000, in response to new scientific findings detecting imidacloprid in maize pollen and confirming high persistency in soils, beekeepers demanded a ban on all uses of imidacloprid. In 2004, the Minister of Agriculture temporarily banned Gaucho® in maize seed-dressing and RégentTS® for all agricultural uses. In 2009, scientific publications proved synergic effects between imidacloprid and Nosema, a small, unicellular parasite or fungus that mainly affects honeybees.

Source: summarised from material in European Environment Agency (2013) *Late Lessons from Early Warnings: Science, Precaution, Innovation*, Copenhagen: EEA.

APPENDIX 2

List of diseases defined as 'neglected' in G-FINDER 2011

HIV/AIDS

Malaria

Tuberculosis

Helminthiases

- roundworm
- hookworm
- whipworm
- strongloides
- elephantiasis
- river blindness
- schistosmiasis (bilharziasis)
- tapeworm
- echinococcosis
- foodborne trematodes / clonorchiasis / opisthorchiasis / fascioliasis / paragonimiasis
- dracunculiasis / guinea-worm disease
- other soil-transmitted helminths

Kinetoplastids

- Chagas' disease
- African sleeping sickness
- leishmaniasis (kala-azar)

Diarrhoeal diseases

- viral diarrhoea
- ecoli diarrhoeal disease
- cholera
- shigellosis
- cryptosporidiosis

Dengue

Bacterial and meningitis infections

- bacterial pneumonia
- bacterial meningitis
- salmonella infections / typhoid
- leprosy

- buruli ulcer
- trachoma
- rheumatic fever
- yaws / endemic syphilis / pinta

Source: Policy Cures (2011) *G-FINDER 2011. Neglected Disease Research and Development: Is Innovation Under Threat?* London: Policy Cures <http://policycures.org/downloads/g-finder_2011.pdf>.

APPENDIX 3

Estimating the costs of true energy access

A study by the Lawrence National Berkley Lab in California investigated the real cost of going beyond improved lighting to provide a reasonably full set of energy services for a low-income household, including electricity for basic appliances and some productive use. The study made a rough calculation of the electricity needs of a low-income rural household to be around 80 kilowatt hours (kWh) per month, based on a daily use shown in the table. Assuming this were produced via a solar photovoltaic mini-grid, the report authors assessed a typical cost at 24 cents per kWh, meaning that level of electricity consumption would cost around $20 per month. This compares to current typical rural household spends on energy (excluding cooking) of $7.50 for 'low-income' households (earning $3–5/day), $4.50 for 'subsistence' households (earning $1–3/day), and $1.50 for 'extremely poor' households (earning less than $1/day) per month (based on the assumption that rural households in developing countries typically spend around 10 per cent of their household budget on energy, half of which is spent on cooking, leaving 5 per cent of household budgets available for non-cooking energy expenditure).

Relative power consumption of appliances useful to low-income rural households

Appliance	*Hours of use per day*	*% of total power requirement*	*Equivalent monthly power use (kWh)*
Medium-sized fan	5	3	2.4
4 LED lights	5	4	3.2
Small TV or ICT device	4	6	4.8
Medium-sized refrigerator	10	29	23.2
Small irrigation pump	4	58	46.4
Total			*80.0*

Source: Kumar, U.J. (ed) (2014) 'Access to electricity', in *50 Breakthroughs: Critical Scientific and Technological Advances Needed for Sustainable Global Development*, pp. 492–525, Berkley, CA: Lawrence Berkeley National Lab.

Index

Aadhaar ID programme, India 80
ActionAid 74, 91–2
advanced market commitment (AMC) 142, 147, 154, 180, 189, 210,
Advanced Research Projects Agency-Energy (ARPA-E), US 21, 178, 184,
agriculture
  extension services 74–7, 86–7, 219, 222
  and ICTs 77, 80–81
  industrialization/Green Revolution 95, 111, 118, 130, 150, 156, 193, 217
  and biodiversity loss 92–100, 137,
  biofuels vs food 94, 120
  livestock 75–7, 92–8
  and planetary boundaries framework 37
  open source seed development 200
  smallholder farmers
  access to technical knowledge 71–7
  funding innovation for 149–156
  traditional knowledge and techniques 93–8, 103–2, 165–6
  agroecology 96–7
  and market driven innovation 143
  Brazil PAIS programme 194, 211
antibiotic misuse and antimicrobial resistance 105–111
  future paths of action 109
  nature of injustice 108
  scale of problem 105
Apple's iPod 183–4
Arocena, R. and Sutz, J. 84, 132

Bangladesh 18, 71–6, 91
banking, mobile 78
biodiversity
  loss and agricultural industrialization 92–8
  and planetary boundaries 26–8
biofuels 91–2
'biopiracy' 165, 167
Bloomberg New Energy Finance 179
Boldrin, M. and Levine, D. 162
Borrás, S. and Edler, J. 232
Boudreau, K.J. and Lakhani, K.R. 203
Bowyer, A. 199
Brazil: Social Technologies Network (STN) 194
Breakthrough Institute 26, 177,
Brundtland Report 25, 60
Buchanan, R.A. 14
Busch, J. 101
business and development sectors: converging views 193–4

capabilities approach 39–40
capabilities development 131–2
Carbon Tracker 98, 104, 224
Castells, M. 84
Chambers, R. 40
Chaminade, C. et al. 132–4
Chesbrough, H. 196
circular economy model 226–8
citizen action and voice 81
clean power-generation, US 176–80
Clements, E. 176
climate change 5–6, 216
  and energy security 98–105
  and food production 111, 118
  and links to poverty eradication 5
  and planetary boundaries framework 28

Climate Technology Centre and Network 240–241
Compton, A. H. 13
computer software *see* open-source software
computer-aided design (CAD): 3D printing 199
computers
history of 15–17
see also information and communications technology (ICT)
cooking practices and facilities 16, 43–6, 49
cost issues
essential medicines 60–6
water and sanitation services 53–5
creative destruction, innovation as 125
Cross, J. 16
culture, influence of 16–17

data analytics (Kaggle) 168–9, 203
data revolution 220–2
digital divide 77–8, 84–5
digital identity 80
disability-adjusted life years (DALYs) 61, 145, 156
d.light 16–7
'doughnut economics' 37–8
drug patents 64–7, 161–9, 170–2

Eco-Patent Commons 170–1
The Economist 175, 184–5
education/training
farmers 7180
open-source innovation 196–7
Ellen MacArthur Foundation 226
energy
access 45–55
and cooking 46
off-grid vs grid 49–52
production and planetary boundaries framework 37
renewables 101–5, 179, 222
sector: venture capital and 'valley of death' in 176–8, 181
security and climate change 98–105
solar 16, 44, 49, 52, 102, 178–81
engagement 182, 191, 193, 222, 231
entrepreneurship 129–32
government role 187–90, 217
environmental movement 25–7
essential medicines 60–6
improving access to 65–6
international trade treaties and intellectual property rights 64–5
WHO list 62–5, 221
ETC Group 73, 95
ethanol, corn production for 92, 118
European Commission: Horizon 2020 initiative 191–2
European Environment Agency 31, 138, 142–3, 145

Fagerberg, J. 127, 130–1, 163
et al. 135, 168
fairness, justice as 38, 91, 109, 123,
Farmer Field Schools 195
feed-in tariffs (FITs) 179
financialization of the private sector 186–7
Fleming, Alexander 105, 118
Food and Agriculture Organization (FAO) 73, 95
Food and Drug Administration (FDA), US 148, 143
food production *see* agriculture
fossil fuels 5–6, 31, 99, 103–4, 137, 150
freedom of choice 39
Fressoli, M.
and Dias, R.
et al. 193, 195
funding
agriculture
advisory services 74–6

innovation 149–54
health research 142–5
responsible research and innovation (RRI) 189–91
see also investment

Gates, Bill 127, 143
Gates Foundation 75, 144, 147–8, 156,
GAVI Alliance 147–8,
Gee, D. 138–41
gender issues 18–19, 33–6, 83–4, 191
agricultural extension services 75–7
equality and Raworth's social foundation 33
mobile phones 83–4
and the EC's Responsible Research and Innovation framework 189
water collection 34, 56–7
General Agreement on Tariffs and Trade (GATT) 164–6
general-purpose technologies (GPTs) 20, 182–3
genetics
diversity/uniformity 92–100
open-source data 200
Global Analysis and Assessment of Sanitation and Drinking-Water (GLAAS) 57
global approach 233
global institutions 233–5
global temperature rise 103
see also climate change
Gosh, S. and Nanda, R. 176–8
governance issues 7–8
broad challenge 231
global approach 233 global institutions 233
single unified problem 118
see also responsible research and innovation (RRI)
government role in innovation systems 132–4, 175–6
entrepreneurial activity 182, 185, 218
rebalancing expectations of public and private sector roles 185–9
venture capital and 'valley of death' in energy sector 176–82
grant financing 148, 167–9s
Green, D. 219–20
Green Revolution *see under* agriculture
greenhouse gas (GHG) emissions 111, 118, 130, 150, 156, 193, 217
see also climate change
GSMA78, 83–5

Hansen, S. and Tickner, J. 139
HapMap project 201
happiness and wellbeing 39
Harvard Business Review 203
health research, open-source 201–2
health research and development (R&D) 142–6
health/healthcare
drug patents 64–7, 161–9
drug patent pools 169–71, 210
energy services required 35
ICT 77
international comparison 62–3
vaccination/immunization 16, 63, 66, 148
'water washed' diseases 53–4
see also antibiotic misuse and antimicrobial resistance; essential medicines
Henry, C. and Stiglitz, J. 163–7
HIV/AIDS 60–1, 164–71
Honey Bee Network (HBN), India 193–4
Human Genome Project 201
hygiene behaviours 54, 59–60

identity/digital identity 80
illiteracy 33, 84
immunization/vaccination 16, 63, 68, 148
inclusion 192–4, 222

inclusive innovation 133, 193, 195–6, 223, 230
business and development sectors: converging views 193–4
India
Aadhaar ID programme 80
agriculture 74, 79, 150–3
ICT/mobile phones 80–4
inclusive innovation 193–5
industry and planetary boundaries framework 37
infectious diseases 53, 61, *64*
information and communications technology (ICT)
agriculture 77–81
Apple's iPod as GPT 183–4
continuing challenge 84
importance for development 77
inequalities 82–4
injustice
agricultural technology use 92–3
antibiotic misuse and antimicrobial resistance 105–11
burden of 215–6
energy access 49–52
water and sanitation access 53–8
innovation 229
global approach 230,
and intellectual property rights/ patents 64–5, 161–172,
justice as fair space for 123–4
policy and drivers 126–32
responsible 190, 229–31
see also inclusive innovation; open-source innovation; responsible research and innovation (RRI)
innovation systems 127–34
and developing economies 130–2
shaping of purpose 142–55
sustainable pro-poor challenge 154
and Technology Justice 210
see also government role in innovation systems; risk management
institutional reflexivity 190
intellectual property rights (IPRs)/ patents 161–4
alternatives to existing system 166–71, 210
drugs 62,65, 163–5, 170
and innovation 64–5, 162–4, 166–8
power asymmetries 165–7
summary and future directions 171–2
TRIPs 64–5, 163–8, 171, 233
see also open-source
interdependency of basic services 87
International Assessment of Agricultural Knowledge, Science and Technology for Development (IAASTD) 96, 150, 152
International Development Research Centre (IDRC) 81, 85
International Energy Agency (IEA) 28, 46–8, 51, 102
International Panel on Climate Change (IPCC) 99–101
international trade treaties and intellectual property rights, essential medicines 64–7
investment
antibiotic development 108
climate change mitigation 134
electricity access 48–50
energy sector 176–187
public and private sector roles and relationship 185–8, 231
subsidy and disinvestment in fossil fuels 104–5
water and sanitation services 53–5
see also funding; risk management

justice
as compromise 32
as fairness 38–9, 127, 223
Rawl's Theory of 38
see also injustice; Technology Justice

Kaggle 168–9, 203
Kanavos, P. et al. 62
Kapczynski, A. et al. 198, 201
Kenya 45, 56, 58, 62, 78–81, 152, 165, 181, 195, 217, 222
  Nairobi 56, 59–60
Khor, M. 165–6
Kim, L. 131
Kinshasa, DRC, energy access 49, 51

Latin America 16, 55, 76, 96, 132, 147, 244
Levy, S. 107, 110–1
lighting (solar energy) 16, 44, 47–9, 87, 102, 178–82
Linux operating system 197
livestock 27, 75–7, 95–9

M-KOPA 180–1
McKinsey & Company 158
markets 6, 19, 91–4, 143–4, 163–4, 175–6184–7
  and patents 163–5
  and pharmaceutical research funding 142–3
  and Technology Justice 218
material aspect of wellbeing 27
Mazzucato, M. 123, 127, 134, 143, 162, 176, 183–7
measurement
  energy access 49
  inadequacy of national statistics 49, 51
  water and sanitation services 57–8
medical research *see* health research
medicines *see* antibiotic misuse and antimicrobial resistance; drug patents; essential medicines; pharmaceutical companies
Mellon, M. and Rissler, J. 141
Mexico: impact of food as biofuel 92, 118
Millennium Development Goals (MDGs) 59, 68–9,86
mobile banking 78
mobile phones 15, 80–1, 87–8, 202
  agricultural market information and trading 79
  digital inequalities 82
  M-KOPA 180–1
  off-grid electricity 180
monsoon floods, Bangladesh 71–2
multilateral development banks (MDBs) 47–8

Nairobi, Kenya 56, 59–60
National Federation for the Conservation of Traditional Seeds and Agricultural Resources (NFCTSAR) 93–4
national innovation systems (NIS) *see* innovation systems
Nehru, Jawaharlal 25
neoclassical economic growth models 124–30
neoliberalism 175
new economies 225–9
New Economics Foundation 231
Nussbaum, M. 28, 32
Nye, D. 3, 15, 18

off-grid/minigrid electricity
  solar lighting 16, 44, 49, 52, 102, 178–81
  vs on-grid electricity 47–50
Open Source Drug Discovery (OSDD) project 202
Open Source Malaria (OSM) project 201–2
Open Source Seed Initiative (OSSI) 200, 209
open-source
  data analytics (Kaggle) 168–9, 203
  definition *196*
open-source innovation 196–8, 204 211, 231
  broader adoption of 204
  motivation in 202–3
  vs open innovation 196–8

open-source software
development 198–9
and licences 198

Oxfam 26, 32, 64, 65, 219
Oxfam America 64

Paris Climate Conference (COP21) 4, 6, 103
participatory technology development (PTD) 77
patent pools 169, 171, 210
'patent thickets' 162, 166, 171, 203
'patent trolls' 162–3, 166
patents *see* intellectual property rights (IPRs)
pay-per-use sanitation facilities 58–9
People's Science Movement (PSM), India 195
Peru 76, 80, 97
pharmaceutical companies 66–7, 144, 164, 169, 171, 183, 201
planetary boundaries framework 37
and safe operating space for humanity 36
and social foundation 32–7
Practical Action 1, 34, 43, 46, 59, 76, 88, 93, 222
and National Federation for the Conservation of Traditional Seeds and Agricultural Resources (NFCTSAR) 93–4
precautionary principle 143–4
predictive modelling solutions (Kaggle) 168
prizes: as incentive for innovation 148, 167–8, 172, 205
public and private sector partnerships 185
roles and relationship 185–9, 231

Rawls, J. 38, 111, 123, 215
Raworth, K. 26, 32–7, 37–8, 71, 82, 91–3, 209
Raymond, E. 197
recycling: circular economy model 226–8
relational aspect of wellbeing 32
renewable energy policy network (REN21) 101
renewables 101–5, 179
RepRap (3D printer) 199–200
responsible research and innovation (RRI) 189–91
applications of 191–2
definition and questions 189
as tool for Technology Justice 192–3, 216
responsive R&D 191
Reuters Market Light (RML) 79
rice: soil salinization problems and solutions 93–5, 102
Rifkin, J. 200, 225–6
risk management 153–5
gaining consensus on 223
risk and reward sharing 187
Rockström, J. et al. 25–7, 32, 34, 37, 92, 111, 209

sanitation *see* water and sanitation services access
Schumacher, E.F. 1–2, 20, 21, 25, 239
Schumpeter, J. 125, 127
science and technology research *see* responsible research and innovation (RRI); risk management
seed development 200
Sen, A. 28, 40
smallholder farmers *see under* agriculture
Smith, J. 15–16, 163
Snow, J. 53
social dimensions of technology 21–4
social foundation 32–7, 38, 95, 119
injustice and technology use 95–6
and Technology Justice 85–9, 218
social impact of technology, unpredictability of 17–28

Social Technologies Network (STN), Brazil 194–5
soil salinization problems and solutions 93–5
solar energy (lighting) 44, 47–9, 87, 102, 178–82
Solow, R. 124–5
South Africa 59, 64–5, 150, 165
South Korea: technological capabilities development 129–31
Sri Lanka: rice cultivation 92–4, 97–8
standard of living 36–40
The State
and the 'Valley of Death' 176–8, 181
as an innovator and entrepreneur 175–80Steffen, W. et al. 25
Stiglitz, J. 169, 163–8, 172, 225
Stilgoe, J. et al. 18, 189–90
The Structural Genomics Consortium (SGC) 201
Sub-Saharan Africa 5, 43, 47, 54, 61, 78, 80, 83, 96, 142, 149, 153–4, 181
energy services 43, *46*, 51
essential medicines 60–3
see also specific countries
subjective aspect of wellbeing 32
Sustainable Development Goals (SDGs) 4–5, 6, 8, 32 86, 103, 119, 134, 170, 192, 210, 219, 234
climate change mitigation 134
and data revolution 220
and social foundation model 37
UN 170, 192, 211, 236–9

Tanzania 58, 81, 180
Tech for Trade 79
technological determinism 15, 17–21, 137, 218
technological momentum, case for 19–21
technological progress, idea of 14–20
technology
defining 13
and human development 3
technology access 215–20
developing world 5, 7
and social foundation model 32–4, 35
and use 119–21
Technology Bank (UN SDG) 6, 86, 170–2–177, 210, 233, 234–5, 238
Technology Executive Committee (UNFCCC) 6, 133, 233
Technology Facilitation Mechanism (UN SDG) 6, 86, 170, 172, 192, 204, 210–11, 234–5
Technology Justice 13, 25–6, 33, 35, 66, 85, 87, 111, 192, 210, 222, 226
defining principle of 38–40
innovation 216–17
responsible research and innovation (RRI) 192–3, 217
and social foundation 85–9, 217
technology transfer/catch-up 124–6, 130–4, 223
technology use 224–5
and access 119–22
10/90 gap 142–3
third industrial revolution 200, 225
3D printing 199–200, 225
Tittonell, P. 149–50,
Torvalds, L. 197
Trade Related Aspects of Intellectual Property Rights (TRIPS) 64–5, 163–8, 171, 233
traditional knowledge and cultivation techniques 93–4, 103–2, 165–6
transparency 221
Treatment Action Campaign 65–6

Uganda 76, 154, 180–1
United Kingdom (UK)
Department for Business Innovation and Skills 134, 176
Department for Culture, Media and Sport 78
patents policy 162

Review on Antimicrobial Resistance 106–7, 114
Sustainability Commission 40
universities: health research 142–3
United Nations (UN) 239–41
Food and Agriculture Organization (FAO) 73, 95–7, 156, 195
Framework Convention on Climate Change (UNFCC) 133, 217, 333–5
Technology Executive Committee (TEC) 137, 233
Global Analysis and Assessment of Sanitation and Drinking-Water (GLAAS) 57, 89
Habitat 56
Human Development Index 39
Sustainable Energy for All (SE4All) initiative 27, 52, 89, 107
Global Tracking Framework 52, 87–9
Technology Bank and Technology Facility Mechanism 170, 223
UNICEF 53, 55, 221
UNICEF/WHO Joint Monitoring Project (JMP) 221
UNIFEM 5
UNITAID 169, 171
Water and WHO 54
United States (US)
Advanced Research Projects Agency – Energy (ARPA-E) 178–9
clean power-generation in 176–9
patents 164, 165–7
urban poor 55–56, 65
urban-rural divide
impact of climate change 102–4
mobile phones 77
water and sanitation services 53–4
Ushahidi (crisis-mapping software) 81–2
vaccination/immunization 16, 63, 66, 148
Valdivia, W. 232
venture capital 162, 176–81, 184
voices
citizen action and 81–2
inclusive innovation and marginalized 193–5

water and sanitation services access 53–60, 90–2
history and nature of injustice 53–6
public investment choices 56–8
universal access 59
urban poor 56
Watkins, A. et al. 128, 132–4
wellbeing 28–35
and standard of living 36–40
Wharton University 92
White, S. 32
women *see* gender issues
World Bank 73, 80, 82, 96, 134, 154, 182, 221
World Health Organization (WHO) 5, 32, 57, 105, 146
antimicrobial resistance 106, 110
global health R&D 151–2, 153
Global Observatory on Health Research and Development 210, 230
list of essential medicines 60–2–7
UN-Water and 56
and UNICEF: Joint Monitoring Project (JMP) 221
World Intellectual Property Organization (WIPO) 161, 164
World Trade Organization (WTO)
TRIPs 64–5, 163–8, 171, 233
and UN, compared 233
World Wildlife Fund (WWF) 25, 27

Zeigler, R. 186
Zimbabwe 76, 81, 165